FROM WAX WINGS TO FLYING DRONES

A VERY UNRELIABLE HISTORY OF AVIATION

NORMAN FERGUSON

Cover illustrations
Front: Aerial Steam Carriage (Library of Congress); Woman riding a horse attached to a balloon (Library of Congress); An accident at Savy Aerodrome, 1918 (Australian War Memorial); Air Force One (Norman Ferguson).
Back: Planespotter on a roof in London (National Archives).

First published 2022

The History Press
97 St George's Place, Cheltenham,
Gloucestershire, GL50 3QB
www.thehistorypress.co.uk

British Library Cataloguing in Publication Data.
A catalogue record for this book is available from the British Library.

ISBN 978 0 7509 9973 1

Typesetting and origination by The History Press
Printed and bound in Great Britain by TJ Books Limited, Padstow, Cornwall.

Trees for Life

CONTENTS

INTRODUCTION

Humankind has always had a fascination with flight. From the first time they could crane their big, hairy, monkey-like heads up, early humans would stare at the skies, mesmerised by the idea of taking to the air, of floating among the clouds and of getting peanuts in silver foil bags rather than having to fight sabre-toothed tigers for them.

Much later, when they'd stopped looking like apes and shed most of the hair, aviation pioneers wanted to emulate the birds. Not the birds who ransacked their rubbish bins or pooped on their charabancs, but those that cartwheeled in the blue skies with abandon, using their wings to pirouette in the free air. (Although some really did want to poop on their neighbours' motor vehicles, especially the ones who played their radiograms too loud on a week night.)

This fascination with getting off the ground – and staying there – would inspire a drive to become aviators, to build machines, to seek out new worlds, to boldly go where ... no, that's *Star Trek*, but it was sort of the same, but on Earth.

Anyway, enough of this. To start at the beginning, there was a Greek person called Icarus ...

ONE

THE VERY EARLY PIONEERS OR I SHOULD HAVE PAID MORE ATTENTION IN WAX-HEATING CLASS

There are many legends and myths from the distant years of yore about humans taking to the air, but one stands out. It is about a young person called Icarus and his dad. Icarus was a Greek person whose dad was another Greek person called Daedalus. One day father and son came up with a plan to escape Crete, where they had been kept captive for some reason or other. Their plan? They would fly out. Easy. One small problem: no airports. Ah. They sat back down and ruminated for a while. Daedalus scratched his chin, hoping that would help his thinking process. It did. He jumped up, exclaiming, 'Eureka!'

'No, my name is Icarus, Dad,' responded his son sulkily, like any teenager would.

'Yes, that's right, son. Well done for remembering. But something even better: I have it!' the dad exclaimed. And he explained how they would build their own airport. Icarus just sighed and shook his head. What a loser Dad is.

His loser dad ruminated some more, scratched his chin some more, and then put his fist onto his forehead (hoping a sculptor would pass by and immortalise his thinking pose). Then he slowly stood up.

'I have a great idea.'

'Oh yeah?'

'Yeah.'

And he did. They didn't need an airport, just aerial flying machines. Being a carpenter, Daedalus was good at making things and knocked up a couple of pairs of wings for him and his sullen offspring. The wings were made of feathers and wax. Being a good dad, Daedalus gave his son some Very Solid Advice: 'Don't fly too close to the sun, as the sun's heat will melt the wax and the feathers will fall out and you will fly like an obese emu carrying a load of house bricks, and then you could very possibly die.'

They launched themselves off into the wild blue yonder, but of course Icarus didn't listen to Pops, did he, and up and up he went, reaching for the skies as only a young man aching for glory but doomed to disaster could. The sun duly melted the wax and down young Icarus plunged, tumbling to the sea below. As he fell, he could only rue the day* he didn't listen to his father. In his head he could hear his old man saying, 'Told you so,' and seconds later he could hear it for real as his trajectory took him past his dad, who shouted, 'Told you so,' at him.

Of course, this is the stuff of legend, and we should place some scepticism around the factual accuracy of this story. For one thing, Daedalus is known as the person who built a maze to imprison a part-human part-bull creature called the Minotaur. Now, we have lost many species over the years due to extinction, but it's unlikely that this mythical monster was ever alive to become extinct. Sorry to be a party pooper but science and its associated facts are things you just have to believe in.

* **That day.**

OTHER DOOMED-TO-FAILS

Icarus wouldn't be the last to disastrously try his hand at flight. In fact, over the years a long list grew of brave and intrepid persons willing to risk injury or worse while attempting to fly.

One such person in the ninth century BAD* was the legendary figure of King Bladud** of the Britons. Now, he might not actually have lived as there's some doubt he was a real person, but we're not going to let facts get in our way. The man who would put the 'dud' into Bladud was said to have discovered the healing powers of mud when the brown stuff rid him of leprosy. More relevant to this tome, however, is that he made himself some artificial wings and tried to fly. He jumped off a temple of Apollo but, unlike the later American space programme of the same name, it didn't result in any giant leaps for humankind. Instead, he made a giant leap for extinction – dying at the bottom of the temple, shortly after being at the top of it.

Another early 'flyer' was Simon the Magician. With a name like that you would expect great and astonishing things. Things not at all like flying around Rome, falling to Earth, breaking both legs and being stoned by a crowd before expiring at the hands of dodgy physicians. Was he magic? Not a lot.

But not all the flyers died. In the ninth century AD, a man called Armen Firman became Armen Airman when

* Before AD.

** An interesting aside – although you'll be the judge of that – is that Bladud's son and heir was a certain King Leir, who later gained fame when the playwright William Shakespeare wrote a play about him called *Hamlear* – no it wasn't! It was *King Lear*, of course.

he jumped off a Córdoba tower in a cloak. Another brave attempt to become airborne was made a few decades after that by another Córdobian called Abbas ibn Firnas. (Although we should say they may be the same person, as back then there was no Córdobian Planespotters' Group around to verify the identification). So when Firman/Firnas covered his brave self with feathers and launched off, he surprised everyone – including himself – by not immediately plummeting to his demise. He glided for quite a distance before coming to a landing. Yes, he was injured, but he wasn't dead. His work in aviation is rewarded with an airport named after him (Ibn Firman/Firnas Airport) and a crater on the Moon. So that's not so bad. It won't pay the bills, but it's a nice-to-have.

Firman/Firnas inspired future jumpers. One was the eleventh century's Ismail ibn Hammad al-Jawhari. IHJ was renowned as a clever man, a bit of a scholar – amazing at quizzes, seemingly. Ask him any country's capital – bam, he knew it right away. But this wasn't to earn him renown. Oh no. One day he announced he was about to fly before a crowd's very eyes. The punters, keen for some entertainment that might involve some risk, stared agog with these very eyes as the brave IHJ climbed up a mosque with a set of wooden wings. The crowd's eyes continued to stare, even more agog, as the brave IHJ made ready. He jumped. And fell straight to his death. No airport for IHJ.

In Britain, undeterred by the very real prospect of failure and death, Elmer had a go. This wasn't the children's colourful elephant character, although he had the same flying characteristics of the species (Dumbo excepted). This was Elmer of Malmesbury, also known as Oliver. He constructed wings and attached them to his limbs. After jumping off the local church tower – the standard operating procedure for instigating flight at this time – this intrepid flying monk actually did a bit of flying. A *bit*. Oliver's armies were getting tiredies and he crashed to the ground, breaking both his leggies and sadly becoming lame for life. Not waiting for the accident investigation board, he immediately blamed himself for not fitting a tail. Oliver later became a prophet, which he might not have been too qualified for, what with the not-seeing-his-crash-a-coming. His work in aviation is marked with a stained-glass window and a pub named in his honour. It's not an airport but not to be sniffed at as tributes go. Pubs are all right.

Despite mixed fortunes, the 'tower-jumpers' continued through the medieval period. One such person was an Italian, Father John Damian, who was part of the royal court of the Scottish King James the Forth Bridge. It was the time of the Renaissance and new-fangled ideas were being discussed hither and thither. New innovations were all the rage, as long as they were approved by the Church. Ask Galileo. Well, you can't, because he long ago entered the Great Departure Lounge in the Sky, but if you could, he'd probably break into a wry smile and say something mysterious in Italian. Very clever man was Galileo.

But back to the courageous Father Damo. Keen to impress, Damian manufactured some wings and, having carefully thought through his next move, jumped off Stirling Castle's ramparts on his way to France. It's good to dream big. He didn't do too badly, only breaking one leg after falling into a dung heap and having onlookers wet themselves laughing.* Father D blamed using feathers of a flightless bird (a chicken) in his wings. Father Damian was also an alchemist, at which he was just as successful, i.e., not at all.

* **It wasn't funny for everyone. The Stirlingshire Planespotters' Group immediately admonished one of their members for not noting down Damian's registration number.**

Another dreamer around this time was João Torto, who lived in Portugal. In 1540 the brave João leapt from a cathedral tower of St Mateus.* As well as his cloth-covered wings, he sported an eagle-shaped helmet. He would have been better donning one resembling a penguin to reflect his aeronautical chances as, when he landed on the chapel roof, his helmet slipped over his eyes and he fell to a not-totally-unexpected fatal death.

They kept coming. In the 1630s Hezârfen Ahmed Çelebi was seen to fly around Istanbul (then Constantinople) using eagle feathers. He is thought to have flown across the Bosphorus. His achievements scared the locals, who rewarded him with a bag of gold and exile to Algeria.

Also in the 1600s, Italian Tito Burattini was another who tried his luck. He was a clever type and had worked out the circumference of the Earth. (It's the distance around it.) In terms of flying, he worked out that eight wings on a 'dragon' would be enough to launch a human. It *was* enough to lift a cat into the air but those aren't humans. Close but no cigar, Tito.

Some of these early pioneers put great thought into their ideas. In 1670 a priest called Francesco Lana published his proposals to build a flyable craft on the notion of four vacuum-filled globes and attaching them to a boat, which would thereby be lifted up. He worked out the size of the globes needed and how to steer while in the air, but the ideas lacked physical existence and Lana's thoughts never got off the ground. Despite this he was called the Father of Aeronautics. And not just by his mum either.

Would a locksmith have the key to personed flight? Frenchperson Besnier the Locksmith certainly thought so. His aviational attempts, unlike his first name, have not been lost

* **Not the patron saint of successful cathedral jumpers.**

to history. He made rods with wings on the end attached to his feet by strings, with which he could 'flap' and so attain flight. He was said to have flown over a house, so that was pretty good, but as a design it wasn't to be repeated. Too much flapping.

Another who was sure he could fly was the Marquis of Bacqueville. In 1742 the bold Jean-François, as he was known to his nearest and dearest, stated his intention to fly across the River Seine using two pairs of home-made wings on his arms and feet. As crowds of undertakers watched expectantly, the brave Marquis jumped off a Parisian hotel's roof. He didn't fall immediately to injury or death but actually managed to get some way across the river before crashing into a boat. And breaking his leg. *Quelle surprise.*

There were some who were keen to progress with flight as long as they weren't directly involved. Pierre Desforges was a clergyperson in France, so he could claim to have assistance from those Higher Up. Would this heavenly help make the difference? Pierre made some wings and tried to force a lowly peasant to use them. The peasant used several Anglo-Saxon words in response and the man of the cloth backed down. He wasn't to give up though. The tenacious D came up with a more elaborate machine. It had a gondola and wings and was duly hauled up to the top of a local community centre. No, it wasn't! It was a local tower. Not finding any volunteers, Pierre went for it and swiftly afterwards reached the ground. He wasn't killed or even seriously injured. Just a broken arm. *Mon dieu!*

Using feathers and lengths of wood hadn't really worked out that well despite centuries of trying, and it was to be another method that brought the success so dearly desired. But before that we almost forgot someone important.

LEONARDO THE VINCI

One of the great minds of his age (he was 37) was an Italian called Leonardo 'da' Vinci. He became so well known he was referred to by his first name, in the same way as musician Madonna Ciccone and jungle survivor Harry Redknapp. Leo was a top-drawer drawer, able to turn his pencil to anything. He could draw a perfect circle without running the pencil around a jam jar lid or anything. Leo was interested in many topics – what we call a 'philomath' or 'clever arse'. In his drawings we see things he drew. One of his most famous was the *Star Jump*, which shows a naked man. (Caution is advised for those of a nervous disposition. He's really naked. You can see *everything*.) Leo could also paint. He did *The Last Supper* and the *Mona Lisa*, the latter of which has been the subject of much discussion over the painting's subject, although there is a strain of thought that suggests she is a woman called Lisa.

But to get back to why we're here: it's the area of aviation that demands our attention. One of Leonardo's sketches shows a flying machine called an 'aerial screw' that, appropriately enough considering its geographical heritage, looks like those twirly pasta shapes. It is the earliest depiction of what we would call 'a helicopter'. Mr da Vinci also came up with the design for what we would call 'a parachute'. This suggests he didn't have too much confidence in the airworthiness of his aerial screw.

TWO

BALLOONING
OR
ARE YOU SURE THEY HAVEN'T INVENTED PARACHUTES YET?

Some clever types figured out that humans didn't have enough power to flap their arms as quickly as a bird. Gyms hadn't been invented and it would be decades until Victorians invented those round barbells that men with hipster beards and waxed moustaches could lift while wearing long johns. As portable power wasn't around, another method of attaining sustainable flight had to be found.

MONTGOLFIER BROTHERS GO BALLONING

It is lost to history just where the concept came from of filling a spherical object with hot air to make it go up. There are reports of inspiration coming from school children inflating frogs and then watching them float over trees, but these have little or no basis in fact.

The French – who definitely didn't blow frogs up (but might eat them) – were first to go down the hot-air route when, in the eighteenth century, the Montgolfier brothers made a flyable ballon.* The two brothers, Monsieur Montgolfier and his brother Monsieur Montgolfier, were paper manufacturers and had made enough money – it definitely did grow on trees for them! – to allow them to follow their dream. Cynical onlookers said they were full of hot air, to which Monsieur Montgolfier responded to huge peals of laughter, 'No, but our ballons are!'

He had a right to say this, as they had invented the hot-air ballon and, in the year of 1783, they put on a public demonstration of their handiwork. They were keen to show their prowess in this awe-inspiring arena of flight, but they weren't daft. The first to take to the air were animals: a sheep, a duck and a chicken. One of these was best placed in case things went awry, having an inbuilt nice woolly cushion, but the other two would have to:

1. Fly to safety
2. Hope its wings evolved into being useful really quickly.

The flight was to take place at Versailles, the royal palace home to no less than the royal persons Marie 'Cake' Antoinette and King 'Louis' the Sixteenth. On the appointed day that would echo down the aeronautical history centuries, the blue touch paper was lit and the ballon rose into the air. *'Oh là là!'*, *'Sacré bleu!'* and *'Zut alors!'* were just some of the stereotypical phrases that might have been said, by those who had learnt their French from *'Allo, 'Allo!*

Up they went into the air and history. The animals flew for eight minutes and then survived the landing, although the chicken was injured by the sheep in its rush to get to the exit door before the seatbelt light had gone off. It was a great success, and the Montgolfier brothers were granted

* **French word for balloon.**

permission by the king to go ahead and risk human necks. There was an awkward silence when it seemed like it might be the brothers' necks, until they quickly suggested that prisoners might be volunteered. This was countered by the argument of 'Who will then make all those mail bags?' It was a good point, said the brothers, running fingers under their by-now sweaty collars. But who would fly on this historic flight?

The two siblings prepared their historic aeronautical object. Their ballon was painted all fancy in blue and red with gold motifs of the sun and fleurs-de-lys* and eagles. They also included depictions of the king's face. As noted before, they weren't daft.

And on a day that will live long in the annals of aviation history, amazed crowds watched agog. Up, up and away went the first aviators in their beautiful ballon. The newly formed Parisian Planespotters' Group excitedly opened their newly bought notebooks and wrote 'Ballon' alongside the date. Undertakers watched expectantly. This new aviation phenomenon looked promising.

The first humans to properly fly rose into the air. The first aeronauts! The Montgolfiers had done it! Ah but wait, that's not the Montgolfiers in the thing. (We did say they weren't daft, didn't we?) Jean-François 'Pilâtre' de Rozier and the Marquis 'François' d'Arlandes were the two who made the first flight and earned their place in the *Côtes de Rhône Book of Records*.** It was they who had climbed into the ballon and watched as the restraining ropes were cut before they could change their minds.

*** Whatever the hell they are.**

**** There are claims that a Jesuit priest in Portugal going by the name of Bartolomeu 'Lourenço' de Gusmão designed and flew a balloon in 1709. If he did, then it knackers up our timeline. Let's just pretend it didn't happen.**

Once up, de Rozier kept the ballon going by stoking the brazier fire. This wasn't a euphemism but how they kept hot air going up into the ballon above by putting straw into the fire. He huffed and he puffed and kept the thing flying.

The ballon floated through the Parisian skies, affording the two flyers a view of the airport R***air would use, far in the distance. They drifted for several miles before coming in for a landing. Part of the ballon had caught fire and it was thought prudent to land before it all burnt through and they fell to their untimely deaths. Undertakers were naturally delighted that these things could catch fire and, although disappointed on the day, resolved to follow the new aeronautical adventures with a keen interest.

They didn't have long to wait. Two years later de Rozier was attempting to cross the English Channel by air. His ballon was part-hydrogen, part-fire source. The ballon's fabric was treated with oil, glue and honey, ensuring pre-flight preparations were constantly interrupted by hungry bears. Once they had been shoo'd off, the flight could commence. Sadly, de Rozier's ballon caught fire and he and his unnamed* companion plummeted to their demise.

A SCOTS AIR

Scotland is the home of many inventions such as television, tarmac and the cloning of sheep. It cannot claim to have invented balloons (although the performance of the national football team at Euro 2020 may give lie to this). However, Scotland does have one claim to fame in the history of ballooning, for on a famous date in 1784 a certain James 'Balloon' Tytler rose aloft over the city of Edinburgh. Citizens

* **Unresearched. - Ed.**

stood in the streets, mouths agape. He wasn't going to 'drop in' for tea, was he?

After a short hop Tytler landed safely and, although he duly earned his place in the *Whisky Tome of World Recoryds* as the first person to successfully fly in Britain, riches and fame were not forthcoming. He etched his place in history in another way. Tytler was the editor of the *Encyclopaedia Britannica* and when you looked up 'aviator' he had written 'Me'. Sadly, the intrepid Scot crashed on another flight. He felt deflated and gave up the ballooning for good. Jamesie didn't receive much attention or cash for his endeavours and, after calling the government of the day 'a bunch of eejits', he was charged with sedition. He soon left for America where, eventually, he died.

THE ITALIAN JOB DONE

As we've seen, flying has always had associated dangers, mostly in the area involving descending too quickly to the ground. Other perils of flight were shown in London in 1784 when Frenchperson Chevalier 'De' Moret had his balloon destroyed by an angry crowd when it failed to lift off. Previous to that, in Bordeaux, another failed ascent led to a riot, with two people being killed and then another two being hanged for their riotous ways. You don't get that at Bristol balloon festival these days. Thankfully.

An ascent that was free of violence took place on a famous date also in 1784, with the first ascent in England. In his balloon, Italian Vincento 'Balloon' Lunardi went off like a shot from London's Artillery Ground. He wasn't alone. His passenger list read like the start of a nursery rhyme as he took a cat, a pigeon and a dog. One had nine lives, one had wings and the other a nervous look on its chops.

The initial part of the flight wasn't without incident. When an oar fell out the balloon's basket, a female onlooker thought it was Lunardi and had a heart attack. The flight continued safely, although the cat, which was male, almost had kittens when the ground rushed up at the end. Balloon landings are rarely as graceful as their take-offs.

Other countries didn't want to be left behind and there were even flights in America, though they hadn't invented it. The Frenchperson Jean 'Pierre' Blanchard demonstrated his aerial prowess in front of the US President, who at the time was George Washington. If Jean had done it later, it would have been a different one.

IT'S A GAS

Hot-air balloons and ballons were all well and good but they suffered from a handicap, which was the actual air inside them. It had to be heated all the time or the whole shebang would simply descend and bump along the ground – if they were lucky. Another gas was required.

This gas was hydrogen, which, as any keen gas student will know, is lighter than air. Not lighter in colour – they're both see-through – but in weight. This means that if you fill a child's balloon with normal air at the temperature of your room, it will just fall to the ground and bump along there as we just mentioned. However, if you take out the air (making sure to produce a noisy 'fart' noise to the guaranteed amusement of all within earshot) and replace it with hydrogen, it will then float upwards and get stuck on the ceiling until Daddy has to risk his neck standing on a chair to retrieve it or face another hour of kid-derived wailing.

Hydrogen was the new 'it' gas – the gas du jour. The first hydrogen ballon* was built by a man called Jacques 'Gas' Charles and two brothers, Monsieur Robert and his brother Monsieur Robert. Confusingly, that was their surname. The brothers had created a new-fangled material of rubberised silk. They used it to make contraceptives and we will leave you to make your own jokes.

The Roberts' design looked more like the balloons we're used to seeing in our skies, with a round gas bag and a basket slung underneath. Like other French ballonists, they weren't daft as their first flight was unpersonned. Their hydrogen-filled creation took to the air and flew for 15 miles before rupturing and falling to the ground, where it was set upon by peasants scared of this aerial monster and who gave it a sound thrashing it wouldn't forget. Or remember, as it was an inanimate object.

The government had to issue a proclamation reassuring the public that not all ballons would fall to the ground and scare them and that, statistically, they were more likely to be hit by a meteorite. This failed to reassure a populace who were now worried about these new-fangled meteorites they'd never heard of before.

Undaunted, *les bold aviateurs* drew lots. Monsieur Robert was chosen and, along with Jacques Charles, took his place in the ballon for The Big One. It duly rose skywards in front of almost half a million people on a famous date in 1783 that would echo down the aeronautical history decades. The undertakers in the crowd went home unhappy, as the gas-filled ballon did not crash. Instead, it disappointingly flew in the sky for a couple of hours. As they passed over the countryside the ballonists could hear the cries of simple country folk below: 'Keep off my land!', 'Off you go, you fancy city types'** and 'Is that a meteorite?'

* **Yes, it was in France.**
** **They didn't say 'types'.**

Among the crowd was a certain Benjamin 'Yes, it's me' Franklin. When he heard another onlooker ask, 'What use is a balloon?' Benji replied, 'Of what use is a new-born baby?' To which the onlooker replied, 'You could win a bonnie baby competition with one.' Benjamin just nodded. You know when someone says something to you and you can't think of a suitable response at the time, but hours later a pithy witticism pops into your head? Benjamin wished that had happened to him, as he walked home feeling outwitted and then couldn't get to sleep for ages.

The ballon travelled 36km and landed safely. Not content with this fine achievement, Monsieur Charles decided to fly off on his own. The ballon's first flight had ended at sunset, but so high did he now ascend that he was able to see the sun again. The brave (or daft) aviator got about 3,000m up but decided to call it a day when his ears got sore. He had jammed his fingers in them to stop having to listen to a loud screaming noise that seemed to be coming from his mouth. Three thousand metres is really high, especially if you're on your own. Charles never flew again.

Showing it was no flash in the pan, Charles and the Roberts designed a ballon that had a rudder for propulsion and steering, but it didn't work. It was not a complete dud though and *La Caroline* went on to fly over 100km, the first ballon to do so. Well done them!

This was a time dubbed 'Balloonmania', a bit like the later 'Beatlemania' but without damp theatre seats. Balloons were all the rage: the beaded car-seat covers and Thunderbirds Tracy Islands of their day. The ballooning triumphs of those mentioned above inspired others to take to the air all over Europe, with some not going down the gaseous bag route. One such person in Belgium, whose name is lost to history and our paltry research, measured dead birds to gauge the wingspan required for his aeronautical device. If that sounds yucky, it is. He built a set of wings and flew a distance of 100m. During the flight it was reported that a blast of wind sent him hurtling towards a well. He was perturbed by this as he wasn't carrying a coin to throw down for good luck. Luckily, the brave Belgian was prevented from falling down it by the breadth of his wings. Lucky duck.

LADIES WHO LAUNCH

On another notable date in 1784 a woman achieved a notable first by being the first woman to fly. The woman was Elizabeth 'No, Not Thimble' Thible and she rose in a ballon made by Jean 'Pierre' Blanchard, who keen-eyed readers will notice appeared back there a bit and will appear in the next bit too. Ms Thible may also have achieved another first by singing opera tunes while in the air. Dressed as a Roman goddess, she rattled off many famous aerial arias – something we are yet to see from Katherine Jenkins.

Another great woman of the air was Jeanne Geneviève 'No, Not Lacrosse' Labrosse, who was married to the celebrated French male balloonist André 'Jacques' Garnerin. André had been the first person to jump out of a balloon and survive using a parachute. There must have been something in the Garnerin water, as Jeanne Genie tried her luck to become the first woman to parachute from a balloon. On her descent from a height of 900m she had time to ponder the meaning of life, and whether her husband had put up those shelves like he said he would. She didn't have long to ponder as she soon came into contact with the ground, the first woman to have made a parachute jump. She'd done it! Jeanne-G also found fame as the first solo woman balloonist, making many ascents and the same number of descents around Europe.

Two other women who flew around this time were Elisa 'Niece' Garnerin (yup, Garnerin's niece) and Sophie 'wife of Jean Pierre' Blanchard. They were rivals in providing entertainment, with night flights featuring fireworks being part of their shtick. Elisa also jumped out of the balloon, an activity that enthralled the crowds agog below, but she didn't die from a too-fast descent – unlike the brave Blanchard, who ruptured her gas bag with a firework and fell to her death, etching her place in the *Pernod Book of Records* by becoming the first woman to die in an aviation accident. It's an achievement you don't really want.

Another notable woman of the air was Mme Poitevin, who made over 570 ascents.* Her party trick was to lift off sitting on a bull or a horse. When she came to London her planned flight with an equine companion was stopped by the police after complaints from the ... ahem ... 'Neigh'-bourhood Watch. You're welcome.

MORE BALLOONS

As keen-eyed geography students will know, the English Channel separates Britain from Europe. To fly over it would be a very notable achievement and Jean 'Pierre' Blanchard (him again) and John 'Jim' Jeffries (first-time appearance) aimed to be those achievers. On a famous date in 1785 that would echo down the decades of aviation history, they left Dover and with the wind in their metaphorical sails they drifted over the coast and beyond. Several times they started descending to the cold, grey, unforgiving water. Thinking quickly, the two bold aviators threw stuff over the side to lighten the load: bags of ballast, packs of leaflets, a barometer, a telescope,

* **571 to be exact. – Ed.**

oars, a kitchen sink unit, their hobnailed boots, two large bar bells, an anvil, Blanchard's collection of horse brasses, a diving helmet and two slabs of marble they hoped to have engraved as souvenirs. Still going down, Jeffries volunteered himself as disposable ballast. Blanchard rubbed his chin: Jeffries was quite a big chap and he was still clutching his complete set of *Encyclopædia Britannica*. But no, it wouldn't be right. Luckily he didn't have time to reconsider the offer as their balloon began to rise. They made it over the coast to the beautiful and solid dry land. Unlike their gas bag, they were not deflated. They were elated. They had done it! Another feather in the French aeronautical cap. Now to find a marble engraver.

Flights like these were very much weather dependent and all these balloons had an issue with steering. They didn't have any. Whatever way the wind was blowing, that's where they were heading. To be proper aero machines they would need to be directed. This led to the development of what are known as 'dirigibles', which means something that is able to be diriged.* And in the nineteenth century they certainly were that, with dirigibles flying hither and thither about the skies.

Pierre 'No Middle Name' Jullien was a watchmaker who looked at the issue of propulsion – not of watches, but of balloons. He did careful work on his theories of using propellers to give movement and direction. It led to him designing a long, thin-shaped balloon that he built in model form. When his fish-shaped balloon pitched up,** he installed a control surface he named a 'fin' at the end to counteract this movement. Jullien's work didn't earn him fame or fortune but perhaps he would have been remembered more if he'd named it a 'pierre' or a 'jullien'.

* **Actually it means 'able to be steered'. – Ed.**

** **Aviation term for pointing upwards.**

Jullien's work inspired another Frenchperson: Henri* 'Steam Engine' Giffard, who in 1852 installed a steam engine in a dirigible. The engine turned a propeller that propelled the craft forwards. It wasn't very good at dealing with crosswinds and, with a steam engine on board, it wasn't superfast: tootling along at 10kmph. This is faster than a brisk walk, similar to that trot/walk/run you do when you're late at the bus stop and the bus is pulling in and you really don't want to miss it but don't want to embarrass yourself in front of the passengers, with everything falling out of your pockets.

But it was the first-ever controlled and powered balloon flight. Giffard received one of the highest awards in France – literally – when his name was inscribed on the Eiffel Tower.

Steam wasn't the only propulsion method thought of though. One flying machine was powered by an electric engine, which was carried aloft with great difficulty as the power cable kept getting stood on and pulled out of the socket. It resulted in a quick redesign, and batteries were installed once someone found that cross-headed screwdriver that opened the compartment. This was the *La France* which, as its name would imply, was built in Spain. No, it wasn't! It was France, of course. On a notable date in 1884 the *La France* took to the air. On board were Charles Renard** and Arthur Krebs.*** Renard, with his thick walrus moustache, and Krebs, with a full-width wax moustache, had done it! Were moustaches a vital element for aeronautical success? The evidence seems incontrovertible.

Steam and electricity were all very well but something more practical was needed. Some kind of engine that could

*** French for 'Henry'.**

**** French for 'fox'.**

***** French for 'something unmentionable oftentimes caught by merchant seamen'.**

combust fuel internally. Engineers and designers sat around pondering how this could be done. Then one of them had an idea. What about an 'external combustion engine'? The cleverer ones pooh-poohed the idea. An external combustion engine was not something that went particularly well with carrying a large amount of flammable hydrogen. But if it was done *internally* then it might just work. If it did, these new-fangled internal combustion engines should generate enough power to propel balloons through the air at speeds approaching 14kmph. Some thought this madness. Undertakers certainly hoped so. Their hopes were helped by the work of German vicar Karl 'God is My Co-pilot' Wolfert in 1897. Karl was first to fit an internally combusting engine on an airship, so etching his name in the *Liebfraumilch Book of Records*. In front of a crowd in Berlin he took off on a summer's evening. Flames were seen to be coming out of the engine's exhausts. More flames were seen to be coming out of the gas bag when the whole thing blew up. Wolfert became Wolfahrt* as he and his engineer then re-entered that book of records as the First to Die in a Petrol-Powered Aviational Craft. At least Wolfert was partly on the way to the Pearly Gates and could expect to go through the fast-track lane.

EVEN MORE BALLOONS

Let's face it, ballooning is exciting. There is always the chance that it might rip, crash into a tree or be punctured by a swarm of angry bees. The public's interest remained high for years and years, aided by direct participation. Showmen would take passengers up (for a fee, of course). In France

* **DYSWWDT? Punning in German, eh.**

in the 1850s the public could be entertained by acts that took to the skies like trapeze artists, ballerinas, ostriches, cows, horses, men on horses, women on horses and carriages getting pulled by horses. 'Yes,' the balloonists cried, 'we need aerial horsepower but this is getting ridiculous!'

Now, as we've seen, there was something about brothers and aviation. Another set of these were the Spencer brothers, big names in ballooning in Britain. Percival, Stanley and Arthur would take ~~innocent victims~~ interested passengers up in the air and give them the chance of dropping back to Earth via parachute. Personal insurance or longevity wasn't a thing back then.

The Spencer brothers had the same father: Charles Green Spencer, so named as his dad had flown with Charles Green. Who he? Charles Green was a British balloonist who brought in new innovations.* One was using coal-gas rather than hydrogen in the gas bags. It was cheaper and didn't involve putting bits of metal into vats of sulphuric acid to get the gas, a procedure that sounds as risky as it is. Green also thought up an ingenious way of controlling the balloon's height. He hung a 'ballast rope' down from the basket that trailed along the ground. When the balloon descended, more of the rope contacted the ground, thus the weight of the balloon was reduced and up she rose! Clever, eh? Green made more than 500 ascents and descents and never lost a passenger. (To be fair the baskets weren't that big.) His flights and inventions helped make ballooning a popular activity. Well done, Charles!

There were also pioneers who wished to expand their scientific knowledge by intrepid means. One was Francisque 'Alps Flyer' Arban, who made the very first flight over the Alps mountain range in 1849. He launched from Marseille

* **Innovations are always new. <Sighs> – Ed.**

(France) and landed successfully in Turin (Italy) the next day. The next month, this experienced pilot gave his wife, Madame 'No First Name Listed' Arban, a flight from Barcelona. That's nice. What wasn't so nice was what happened next. He dropped her off and took off before disappearing over the sea, never to be seen again.

Of course, it would be amiss to leave out the flight of James 'Double' Glaisher and Henry 'Balloonist' Coxwell. Glaisher was big into the weather and has been described as the father of meteorology, which seems misleading because, as far as we know, meteors have existed in space for ages. But anyway, what do we know?* Glaisher wanted to know more about what the sky was like, and the only way to get there was via balloon. On a famous date in 1862, in a coal-gas-filled balloon, the two rose. And rose. And rose. It's not clear how high they got – both aeronauts had their eyes firmly wedged shut at this point – but it's reckoned around 37,000ft. So that's like going up Everest and then a couple of Scottish Munros on top of that. They had no oxygen with them or enough clothes. Glaisher passed out due to the lack of oxygen, or it might have been realising they were 7 miles up above the lovely, warm, safe ground. Coxwell's hands froze and he had to open the gas valve with his teeth. Nice party trick, Henry. This allowed the balloon to start descending. On his return, Glaisher announced his findings: 'It's very cold up there. Very cold.' He also found that the wind direction varied at different heights. So that was interesting. He also never flew again.

In France, similar flights ventured to places Very High Up. In 1875 the *Zenith* took a crew of three to 28,000ft but sadly didn't bring them all down. Well, it did, it's just that they weren't all alive. Two had died through lack of oxygen. At one point

* **Very little? – Ed.**

a crew member had passed out, come to and in his confusion dumped ballast. Not a good thing to do. After this, such flights were deemed 'bloody dangerous' by the risk assessors.

Another great and epic flight was that attempted by Salomon 'Auguste' Andrée. He was going to be the first to fly over the North Pole in a balloon, which he called *Oern*. Sadly, the wisdom of Salomon was not in balloon flying. He didn't *Oern* on the side of caution* and on a date in 1897 Andrée took off with two others. Nothing more was heard from them for three decades until their bodies were found. Their journal and undeveloped films were recovered and showed how they had flown for over sixty hours but then had to come back via a ground route and perished some months later. This section is a bit gloomy, isn't it? It doesn't get any lighter in tone for a bit. Sorry.

One epic flight that was successfully completed was that by American balloonist John 'Ernie' Wise. Wise by name and also by nature, he devised a method by which the balloon could be deflated. He installed a cord that, when pulled, said it was a failure and would never amount to much. More useful for aviation, it also opened a portion of the balloon, thus releasing the gas inside. It's a bit like when a school child asks you to pull their finger.

Wise was keen to travel long distances; whether or not he got on with his family isn't known. He wanted to establish transatlantic balloon flights and in 1859 carried out a very long flight of over 800 miles, from St Louis to New York state. His balloon carried the first ever airmail. So that's good. Wise had competition from a man with a perfect nineteenth-century name: Thaddeus 'C' Lowe. Both wished to be first across the pond, but unfortunately Wise couldn't cross Lake Michigan, dying in a balloon there in 1879.

* **<Groans> – Ed.**

Balloons could be used for testing experimental craft. In 1874 a bold Belgian called Vincent 'de' Groof became Vincent de Ooft when testing a 'flying' machine he had devised. It was a vertical frame in which the ~~doomed~~ brave aeronaut stood. Above him was a set of hinged, flapping wings that would provide lift. On release from the balloon the contraption folded up and de Groof fell to his demise, exchanging his wooden wings for a pair of angel's. His balloonist, Joseph 'Balloonist' Simmons, almost came to grief, as the lightening of the load caused a rapid ascent, during which he conked out from lack of oxygen. He came to with his descending balloon about to land in front of a steam train. His luck was in this time but ran out when he died in a – guess what? – balloon accident in 1888.

Simmons wouldn't be the last to perish from aeronauting: a situation that pleased funeral directors, although it wasn't the case that Parisians had to wear hard hats to avoid the perpetual rain of falling ballonists. Most flights were carried out perfectly safely. Honestly.

ARMY AIR FORCES

It crossed the minds of some that balloons could be used for military purposes. These minds were French, and they had created a balloon corps that was used for reconnoitring enemy positions, watching troop movements, smelling what they were having for brunch – that sort of thing. A balloon was first used in battle in 1794 at the Battle of Fleurus, where Austrian troops responded in a fairly typical manner by being very frightened of the *l'Entreprenant* as it floated over them. Unfortunately, Napoleon 'Emperor' Bonaparte abandoned his balloon plans before they could have helped him when he met his Waterloo. At Waterloo.

Balloons offered other opportunities. An enthusiastic German called Franz 'Keen' Leppich came up with the idea of a balloon war machine. He tried to flog his idea, but it was rejected by Napoleon and the King of Württemberg (wherever the hell that is) before being picked up by the Russians, who figured it could be used to drop blowy-uppy things on an enemy. The Russians had 'defeating an enemy' high up on their to-do list as in that year – 1812 – Napoleon was making overtures on their country. A wonder weapon would be great at stopping the French military genius (and all the troops that came with him) before he reached Moscow. Unfortunately for the Russians it all went pear-shaped for the pear-shaped machine as it didn't really work.

The French weren't done with their own military balloons and planned to deploy one at the siege of Algiers in 1830, although there is nothing on Wikipedia that says they did. In 1849 there was still war balloon interest. The Austrians were besieging Venice (during tourist season it's the only way to get in). Unable to achieve access, they launched a hundred balloons carrying bombs with timing devices. The raid had a 1 per cent success rate, with only one exploding in the target city. Most of these new-fangled weapons were affected by wind and blew back before exploding *inside* the Austrian camp. You had *one* job.

Later, during the siege of Paris (again, a very busy tourist spot), balloons were used to spot enemy movements and also for carrying people and mail out of the beleaguered city. Many postcards were inscribed 'Not having a good time, don't wish you were here'. Unfortunately, not all the balloons came down where they were intended to, with one balloon landing in Norway, which is quite far away. More than 2.5 million letters were transported out of the city, with carrier pigeons being used to return messages. Most got through, with the besiegers' special aerial unit of assorted flying machines unable to *Arrêtez les Pigeons*.

Other countries' military commanders didn't want to miss out. Italy, Spain, Switzerland and China were four of them. In America, balloons were used in the American Uncivil War. In Britain it took a bit longer – the next century – until the army established its accurate, if uninspiring, Army Balloon Equipment Store in a shed at Woolwich Arsenal. A few years later the just-as-prosaic Balloon Factory was created, along with a Balloon School to train Balloon Pilots. The name of the man who took command is familiar to fans of the time-travelling alien show *Doctor Who*. Colonel James Lethbridge 'Tim' Brooke Templer was the man with the right skills, experience and moustache to take military ballooning forwards. And upwards. The army balloonists went to Africa, not for the wonderful scenery, but to take part in the Boer War, where they were able to spot the Boer soldiers easily from the air. This success meant bigger facilities were required and so the move to Farnborough was made. Interestingly, artillery guns were used to remove trees growing on the field, an option that sadly isn't open to everyone wishing to tidy up their garden.

Those based at the new Balloon Factory could enjoy working in the Airship Shed, the Kite Room and the Skin Room, which might sound unpleasant, because that's what it was. The balloons' skins used cows' stomachs and they were prepared on site. It was a vegan's nightmare and you can only imagine the stink.

SANTOS IS COMING TO TOWN

Around this time a figure appeared who was to go down in aviation history. His name was Alberto 'Santos' Dumont and he was a moustached Brazilian. His family had money and he spent a lot of it on flying. He lived in France – well, why wouldn't you? – and could be seen flying above it in small airships. Or *into* it when, on his first dirigible flight, he steered straight into some trees after take-off. Alberto didn't die though and went on to build others (dirigibles, not trees). His *Number 3* flew at a stupendous 12mph. Alberto was a popular figure as he puttered over Paris in his bowler hat, descending for luncheon appointments and parking his ballon nearby. *So* French. And Brazilian.

As would become the fashion, competitions were a way of getting things moving, and a millionaire called Henri 'Deutsch' de la Meurthe put up 100,000 francs for the first aviator to fly from Parc Saint Cloud, around the Eiffel Tower and then back in under half an hour. On a famous date in 1901 the bold Alberto took to the air in his *Number 5* airship – and landed on top of a Parisian hotel. He had lost gas – it was quite scary up there – and had to descend. The bold Brazilian was rescued from the hotel roof by the fire brigade: Pierre, Pierre, Barneau Magreaux, Cuthberthe, d'Ibble et Grubbe.

A few months later Alberto tried again. In his *Number 6* airship, he took off, headed to the heart of the great city and flew around the Eiffel Tower, a place he'd always wanted to visit but baulked at the expensive admission prices. However, there was a slight problem. His epic flight finished forty seconds outside the time limit.

'Awww,' said Alberto, 'give me a break.'

'No luck,' said the judges.

'Give him the d'argent!' said the crowd and, knowing what can happen with a rowdy French crowd in Paris, the judges caved and popular acclaim won the day. Alberto got his money and instead of buying fur coats, gold sovereign rings and flash motors, he shared it out among the deserving poor. The undeserving poor got some, too. The Brazilian government matched the prize and gave him the same amount. Alberto gave that away. Then they matched that generosity. And he gave that away. It went on like this until officials were stuffing banknotes into his pockets like arguing aunties at a cafe.

It was time for more brothers to make an appearance and Monsieur 'I Ain't Got' Lebaudy and Monsieur 'I Also Ain't Got' Lebaudy started building semi-rigid dirigibles, which is not an easy thing to say after a couple of Pernods, which is what it would take to gain enough French courage to fly in one. However, semi-rigid dirigibles were not going to set the pace. It was time for the German airship to sail into history.

LEAD ZEPPELINS

One man who was to etch his name in the *Schnapps Book of Records* was Count von Ferdinand von Graf von Zeppelin, who wanted to bring new innovations into the world of flight. He was suitably equipped with a keen mind and, more importantly, a terrific walrus moustache. Few aviators of the period would want to be seen climbing out of their wreckage without one, and Ferdy had a corker.

As those with a reasonable memory will remember, Germany was the first country to have an airship fitted with a petrol-engined motor. Yes, it had blown up but when did that put anyone off? Zeppelin was forward-looking (not just due to the placement of his eyes on the front of his head) and he thought large airships could offer a new mode of transportation that could carry people relatively safely over the Earth. His airships were 'rigid dirigibles', constructed with an internal metal frame, inside which the individual gas bags were very carefully placed. The engines, as well as the crew and passenger areas were slung underneath. The paying passenger could look down or across but not really up. This was probably a good idea, as a constant reminder of those gas bags full of highly explosive material might mean it took a wee bit longer to settle down to sleep at night.

Zeppo's first flying example was given the suitable name Airship Zeppelin One in 1900. To give some idea of its size, it was 420ft in length. On the first flight it almost went into a lake, but quick thinking and reverse thrust prevented it going down like a ... lead balloon. Airship Zeppelin One was succeeded by Airship Zeppelin Two, Airship Zeppelin Three and so on. A sensible naming system and no mistake.

In 1908 Zeppelin suffered a setback when Airship Zeppelin Four was destroyed by earth, wind and fire. While moored on the ground, wind pushed it into some trees, which tore open a gas bag, which let out the hydrogen, which ignited with a stray spark.

The population clubbed together, and public subscription raised money for a new one. Nowadays we get such public fundraising to buy someone a new puppy or replace stuff they lost in a lightning strike; back then they built airships.* The Airship Zeppelin Five flew fine but on one of its jaunts it collided with a pear tree, damaging the front bit. The enterprising crew just hacked off this piece and made a temporary front section before taking off again. The incident took place in May but if it was nearer Yuletide then someone might have made a bad pun about a part-rigible in a pear tree.**

These Zeppelin airships were seen as the future of air travel. You could drink cognac, eat all the bratwurst you could eat and exchange other German stereotypes in perfect ease as you floated over the Atlantic, and if you'd bought

* **Puppies are cute though. - Ed.**

** **God help us. - Ed.**

a return ticket, you could do the same on the way back. Passengers were afforded great luxury and could gaze out the windows upon galloping giraffes, playful penguins and lying-down lions, and once they'd flown past Berlin Zoo, they could look down on other things.

The brand identity of Zeppelins had been tarnished by them being used during the First World War to drop bombs on people and then to carry the flag of the Nazis around the place but, while this was bad, something badder was to deflate the airship bag for good.

RISE AND FALL OF THE *HINDENBURG*

The *Hindenburg* was the biggest airship in the world. At over 800ft long, it would be the biggest thing *ever* to fly. Well, that and its sister ship, *Graf Zeppelin Two*, which was the same size. The Nazis were very proud of their big airships, but you know that expression, 'Pride comes before an explosion'? Many airships had come to disastrous ends* but the *Hindenburg* was to claim top spot – oops, belated spoiler alert. In the Top Trumps *Aviation Disasters* deck it scores very highly. In a strange omen of what was to come, they had used about 5 tons of metal from the British R101 in making the *Hindenburg*. The R101 had crashed and burst into flames on its first voyage, killing most of its passengers and crew. Not exactly a rabbit's foot when it comes to conjuring up good fortune.

In 1937 the *Hindenburg* had flown across the Atlantic from Germany without exploding and it was hoped this would

* **R38, R101, USS *Akron*, USS *Macon*, USS *Shenandoah*, *Dixmude*, LZ 14, LZ 18, LZ 40, LZ 54, LZ 104 and LZ 112, to name twelve.**

continue. Zeppelins had carried passengers for thousands of miles without any copping it, so why should this flight be any different?

As it approached the landing field on an infamous date, all eyes were on the huge silver Zeppelin-shaped Zeppelin. Some of the onlookers were agog. Some were awed. Some were just thinking of what to have for dinner.

This huge aerial behemoth was quite a sight as it slowly came in. At this point we should probably mention that the aerial giant was filled with hydrogen and not the much safer helium, as the USA wouldn't sell this safer gas to Germany. This is quite important as … suddenly fire was spotted and it quickly spread along the fuselage, burning away the covering, revealing the doomed machine's bare metal frame. Panic ensued, with news cameramen worried they were going to miss this scene of destruction and death. An American broadcaster became famous for his impassioned commentary on the horrible scene: 'Look at that airship, it's on fire. It's terrible. The humanity. Have you thought about what kind of insurance policy you need in later life? Why not try Premium Care Insurers? They Care About You.'

The fire killed thirty-six people, with some crew and passengers able to escape from the towering inferno that lit up the sky. Passengers who had thought about taking an airship trip looked at each other, wrinkled their noses and said, 'Nah.' Airships were done. Yes, they were nice to look at but if people were going to die in a fiery furnace, they wanted to do it in much faster machines. The aeroplane was ready to take its place.

THREE

POWERED AEROPLANES OR THE WRIGHT PERSONS FOR THE JOB

What was it with brothers and early aviation? We just don't know. Maybe they got bored fighting over whose turn it was next on the rocking horse? Whatever the reason, they produced the aviational goods. In Ohio, America, two young American brothers: Mr Wright and his brother, Mr Wright, were keen to make inroads into the world of flight. They made bicycles for a living and gained important insights into areas such as precision engineering and tucking your trouser legs into your socks to stop oil getting on your expensive pants. They had a slight disadvantage though: only one had a moustache.*

At first, the dynamic duo built kites. Not the ones that get their cables all twisted and lie neglected in cupboards after one session at a park, but big kites, with two wings mounted on top of each other – sort of like bunk beds, but bunk beds that could fly.

* American word for trousers.

These early kites gave way to bigger versions that were carefully designed to carry a ~~complete lunatic~~ bold aviator. These were called 'gliders' as they were designed to 'glide' through the air. It was a perfect description – if they worked. If they didn't, they could be called 'plummeters' but these Wright brothers knew what they were doing. At one point Mr Wright was asked if his glider was strong enough to fly his brother as well. He responded in a way that made future songwriters reach for their notebooks, 'He's not heavy, he's my brother, and the load-bearing factors of the fuselage construction have taken that into consideration.'

These brothers of the same mother (and father) went further. They made a glider that could be steered in the air. Now this might sound pretty obvious, but before this, glider pilots would launch themselves off into the blue yonder and then fly in a straight line (barring crashes, of course). This straight-ahead-only flightpath had obvious limitations. But the Wrights realised if they 'warped' the 'wings', they could induce the 'aircraft' to 'turn'. They added an 'elevator' to move the aircraft 'up' or 'down' and a 'rudder' at the back to decrease what the aero-boffins call 'adverse yaw', whatever the 'hell' that means.

This was exciting but it wasn't going to get them to Acapulco for the weekend, so they thought about putting an engine on their aircraft. Then they did it.

On a date in 1903 that will echo for a long time in the annals of aviation history at a charmingly named place called Kill Devil Hills in a state in America, the two brothers made their giant leap into the history books and skies. Before their famous flight the two brothers had not decided who was to be the intrepid First to Fly. In time-honoured fashion they tossed a coin. Mr Wright won. He asked for it to be 'best of three' but his sibling wasn't having it. He duly took his place in the aircraft, which had been given the slightly presumptuous name *Flyer*. The engine was started. The propellers were spun. Those watching stood agog. The long-winged craft wobbled with power and maybe nerves. The tension rose. Then Mr Wright remembered to let off the handbrake. Off he went! The Flyer zoomed down its launch rail and took to the air, reaching the dizzying height of 15ft before descending rapidly in what is often described as a 'crash'. The shaken pilot got out, glad he had worn the brown cords, and the chance of being the first to fly a sustained flight passed to his brother.

The other Mr Wright prepared himself for glory. The brothers' two-winged aero machine puttered down the launch rail and then, magically, beautifully, but also very slowly, rose into the air. It flew! It didn't crash! A photographer was there to capture the moment that would appear on book covers, mouse mats and tea towels under the appropriate copyright licensing arrangement for decades after. The *Budweiser Book of Records* would state that the *Flyer* covered 120ft in four seconds, which means that its average speed was ... you take the big number and multiply, no, divide by the number you've thought of and ... it doesn't matter. The main thing was, them good old Wright brother boys had done it.

As they whooped and hollered, the local undertaker turned away, regretting coming out all this way to see nothing, while the newly formed Kitty Hawk Planespotters' Group happily wrote 'Wright Flyer' in their notebooks.

The world had changed. With this simple, short and rather slow flight, humankind had followed the birds into the air. Soon everyone would be building their own flying machines, called aeroplanes in Britain and airplanes in America. In France, just to be different, they called them *avions*. In Germany they went further, calling them *flugzeugen*. Whatever they were called doesn't really matter – we just like looking things up on the internet.

The Wrights did not rest on their laurels. (They had kept details of their machine quiet and so didn't actually get any laurels.) The laurel-free brothers continued to develop their craft – in both senses of the word – and improved the engine's power, put in a seat and gave their airplane the ability to turn, so that it could make its way back to where it had started from. Clever guys, these two. They clearly had the Wright Stuff.

Those of a military mind could see the potential of these new-fangled airplanes. In America in 1907 the Wrights were asked to demonstrate theirs to military commanders, as the president wanted an aircraft that could fly him around and save having to travel economy. The brothers were ready. They went to Washington to showcase their aircraft's potential as a military machine. It certainly showed its capabilities as a killing machine as, during a demo flight, it crashed, killing an army lieutenant – the first passenger to die in a plane crash. It's not an achievement you'd want in the *Bourbon Book of Records*, really.

Now you might think the military would baulk at progressing any further, as machines that kill their own personnel are not a hugely attractive sales proposition. But generals are not always put off by the deaths of their men, so they continued wanting a piece of this airplane action.

They wouldn't be the only ones.

HIGHLAND CAYLEY

No account of aviation should forget (like we almost did) to include the man known as 'the father of aviation': Sir George 'Father of Aviation' Cayley, from Britain. He was born in 1773, which keen-eyed arithmetic fans will know is 130 years before 1903. Cayley inherited lots of money and huge tracts of land.

With plenty of time and money on his hands, he was able to indulge his interests. It wasn't stamp collecting or brass rubbing for Georgie Boy. Oh no. He invented seat belts, a way of draining fields, flotation tanks for boats, artificial hands for amputees, fountain pens and, most importantly to us, the aeroplane.

Yes, Cayley was the first to write down the outline of what we would recognise as an aircraft, i.e. a machine with wings to keep it up, facilities to steer it around, a place for a person to sit and a method of propelling it through the air. He looked at the things that could ensure flight, things like lift and drag. He thought about how streamlining would help reduce this drag and how a curved wing made more lift than a flat one. For propulsion he came up with the idea of 'flappers' but quite how he thought dancers from a future decade were going to move an aircraft through the air is sadly lost to history. He needed more power, Captain, but at the time, there weren't engines small enough and powerful enough to fit onto something that could conceivably climb into the sky.

So, Cayley built gliders, the first one flying in 1804. He dabbled with balloons – they are a lot of fun – but came back to winged machines. He built one with three wings called the Boy Carrier and it did exactly that, carrying a 10-year-old boy off the ground in 1849. This was the first-ever heavier-than-air flying machine to ever fly someone. How about that, eh? Not bad going. The boy's comments are not recorded but it is thought the swear box emptied his piggy bank.

Cayley then designed a flying machine called the Governable Parachute, which looked like a large flat fish positioned over a wheeled boat that has a bird's head on the bow and a rudder sticking out the back. In 1853 Cayley was to reach his highest achievement, even though he was firmly on the ground. His glider, also known as a Man Carrier

– you know where this is going – was tested and, yup, it flew, travelling 900ft across the green and pleasant land. Aboard was Cayley's coachperson who, when back on the Earth after his epic flight, resigned on the spot, saying, 'You can stick your job up your arse, Mister Cayley.' The name of the pilot of the first-ever flight by an adult human in a flying machine is lost to history and can't echo down the decades.

Cayley isn't a household name, although this could be down to the lack of a moustache and a poor PR company. His work wasn't for nothing, as it influenced the Wright brothers, who said nice things about him. Cayley had calculated that thrust is needed to overcome drag, and lift is needed to overcome gravity. Cayley did this in the nineteenth century, so you do wonder why others spent so much time faffing about with balloons, but faff about they did. There was a rumour that he eventually did manage to find an engine small enough to go into one of his designs but whether he was the first to design a powered flying machine a century before the Wrights we will never know, unless time machines are invented.

Cayley's contribution to aviation history is marked with two student halls of residence named after him, at Loughborough and Hull. So that's something. It's not a big bronze statue, but it's better than nothing.

FULL STEAM CARRIAGE AHEAD

In the nineteenth century it was said that coal was king. Not only is this gender-determinant, it is also incorrect, as Queen Victoria was mostly king during this time. Anyway, we're getting diverted. Coal produced steam in steam engines, and this could be used to power things like trains, ships and machines in factories. Could this steam power aircraft?

William 'Steaming' Henson and John 'Also Steaming' Stringfellow had high hopes for the water vapour-producing energy source. Their background in the lace industry might not suggest an immediate advantage in aeronautical advancement but they both sported fine moustaches (and sideburns), so they weren't without a chance of glory.

Together they devised the Aerial Steam Carriage. It was going to have two wings, a fan tail and a short fuselage with space for a steam engine. It was gonna be big: 160ft-wingspan big. Its propellers would be 23ft in diameter. It would carry passengers at 50mph. It would carry these passengers majestically over the Pyramids. *Would* is the key word here as it never flew. Sadly, the pair couldn't raise enough money and so the carriage didn't go anywhere, unlike the fed-up Henson, who left for America, never to dabble in flight again. However, his work in aviational development saw an Antarctic glacier named after him in tribute. It's not a student hall of accommodation but it's something. Apropos of nothing, he invented the T-shaped razor. Thanks, William.

Stringfellow, though, still had the flying bug and designed a flying machine with a 20ft wingspan. He took it out to an area of the English countryside near Chard,* where it refused to become airborne. For seven weeks Stringfellow encouraged, cajoled and thrashed his 'flying' machine with a stick but it didn't comply. It was just too heavy and too underpowered. Disappointed, he trudged back home. But what if he made it lighter and smaller?

The next aircraft had a wingspan of 10ft and was fitted with a less-heavy steam engine that turned two propellers. On a famous date in 1848 this monoplane – called the Bat because of its wing shape and fondness for insects – flew. Even though it was inside a mill, it was the very first powered

* **No idea.**

flight. Those watching celebrated with a cuppa – the flying machine's power source was doubly useful.

Stringfellow got a bit bored and went off to do other things but the bug bit again in the 1860s when he heard a talk by a person called Herbert Wenham called 'Aerial Locomotion', which sounds like he would be talking about trains in the sky (it's one way of avoiding those pesky leaves) but it wasn't. It was about how to fly. Stringfellow was inspired and produced a new design. His craft was a triplane with square wings and a big not-square wing at the front, or possibly at the back. It's difficult to tell from the photo we are looking at. It was said to weigh the same as a goose. A big goose or small goose? We just don't know. It was shown at Crystal Palace in London, where the model was given a length of wire to fly down, as the organisers were worried it might start a conflagration* if allowed to fly free. Seems sensible.

Stringfellow wanted to build an aircraft big enough to carry a human but he was in his eighties by then and these were the days when your eighties were not the new fifties – they were the eighties. He died in these eighties, in 1883. His work was marked with a bronze model of his flying machine in Chard town centre and a stamp in Cambodia. First class.

Around this time, there were other attempts, too numerous to research, to devise a working winged flying machine. A Frenchperson called Alphonse 'Inventor' Pénaud came up with a method of propulsion that was to endear him to millions of school kids through the ensuing decades: the wind-up rubber band. When twisted, the kinetic energy stored up in the band would power a propeller and drive his lightweight models through the air. *C'est voilà!*

In the 1890s Sir Hiram 'And Fire 'em' Maxim built a flying machine that had a couple of issues:

* **Posh word for a fire.**

1. It weighed 4 tons, and
2. It couldn't fly.

To be fair to Hiram, he did say that it wasn't meant to fly and was just a test rig. It had two very large propellers, powered by two steam engines, and its construction is hard to describe. The simplest way is to imagine a huge thing of material and steel tubes all connected by wires and struts and something you might shelter under at a summer party. The gigantic machine with its 100ft-plus wingspan was built to run along tracks but during one test it produced enough lift to lift off. Yay! In doing so it pulled up the track designed not to let it fly and it crashed. Not yay. Sadly for the watching funeral directors, no pilots were harmed in the making of this crash. They were much happier with his other invention: the Maxim machine gun, used at places like the Somme and Passchendaele.

Hiram did not waste his expertise in flying machines that didn't fly. He built Captive Flying Machines: amusement rides whereby gondolas were flung around a central structure and which have been convincing children they should have gone to the toilet first for many years now.

FLAPPING AROUND

The word 'ornithopter' comes from Latin, meaning an 'orni' that 'thopters', and were contraptions designed to mimic birds' wing-flapping actions. The thirteenth-century English monk Roger 'Friar' Bacon wrote about them and, of course, brainbox Leonardo 'da' Vinci designed one. They were seen as the way forward for centuries until technology and sense was seen in form of the balloon. There was a major problem with the ornithopter: it was difficult to get the wings flapping quickly enough to gain enough lift. A human's arms just aren't strong enough to do this, even someone who works out or lives on their own. Da Vinci suggested mechanical means could be used to get the wings flapping. And he was really clever, so it wouldn't pay to pooh-pooh his pondering. It might just work and, in the nineteenth century, it did. Ornithopters actually flew. They actually did. In 1890, Frenchperson Gustave 'Cartridge' Trouve used pistol cartridges to push the wing down, which then returned upwards by use of springs. It was claimed this ~~bizarre contraption~~ amazing flying machine could climb to a height of 200ft. It should be said that Gustave's machine wasn't carrying a human, but it was still good.

The first ornithopter to fly with a person in it was designed by Alexander 'Not on the' Lippisch. Alex generously allowed a pilot called Hans 'Werner' Krause to risk his neck in it. The 'thopter was propelled into the air by an elasticated cord and this led to scepticism by sceptics, who felt it was only gliding. 'What's the point in this?' asked many, who felt aircraft with engines were an all-round better way of flying and not this larking about. There were later attempts to fly flapping-wing aircraft but we're with the let's-not-lark-about faction so will ignore them.

One thing we couldn't ignore, though, was the excitement caused by the arrival in Europe of two successful American brothers.

FOUR

THOSE MAGNIFICENT PERSONS IN THEIR FLYING MACHINES OR A LEATHER CAP DOESN'T OFFER MUCH PROTECTION, DOES IT?

In Europe, when they heard of the Wright brothers' success, they smashed glasses, kicked over waste paper bins and raged at the skies, clenching their fists, yelling words that started with 'f' that were not 'fantastique', 'fantastisch' or 'fantástico'. They were not going to let those Americans over the Atlantic win the skies unchallenged. 'N'pas mon amis!' said the French, who were particularly advanced when it came to early aviation. Men like Louis Bréguet, Louis Blériot and – guess what? – yes, more brothers, like Monsieur Voisin and his brother, Monsieur Voisin. And Monsieur Caudron and his brother, Monsieur Caudron. They were determined to master the skies and show those Americans what was what.

They were going to get a chance to do this in person, as in 1908 the Wrights made the journey over the Atlantic to Europe. Not in an aircraft – they weren't *that* good – but in a boat. They had brought their aeroplane with them, though, which they reassembled, only having a couple of wingnuts left over from the packaging.

At a packed air show near Le Mans (The Men) they took to the air. The expectant crowd hushed as the Wright Flyer lifted off. Some were agog. A couple were awed. Then, before their very eyes, Mr Wright performed something that defied logic. He turned the plane *while still in the air* and set it off in a different direction. The French remained stunned, motionless, as if turned to stone. If a French aviator wanted to turn their plane they had to land, push its nose towards the direction wanted and then take off again. It wasn't particularly effective or efficient, and when they saw these brash Americans blatantly shoving their prowess down their baguette-chutes, they burst into tears of impotent rage. Europe was not as ahead as it thought.

CHANNEL DASH

While France was behind America, it was in front of a certain country that rhymes with 'Gate Bitten'. In 1909 an event took place that made the teacups of British military circles tinkle in apprehension. For in this year a human did something no one had ever done before: they flew across the English Channel in an aircraft. Not only that, but they did it deliberately.

Louis 'Aviator' Blériot was a bon viveur, a raconteur and an entrepreneur. Yes, he was French.* More importantly, he had a fine moustache. This aviator was part of the French pursuit of flying that was blazing a trail (and unfortunately a lot of pilots who weren't wearing flame-proof pants) in aerodynamic innovations. For example, Blériot used rice paper on some of his aircraft rather than fabric. It was a good material for covering the fuselage, although the drawback was having to fight off the squirrels who had acquired a taste for it.

* **We know. You mentioned it just above. - Ed.**

Blériot sold car accessories, and when it came to this new-fangled aviation he was not going to be standing around like a spare part. He was in competition with Hubert 'Pioneer' Latham, who was a fellow French pioneer. Latham had no moustache but was a brave aviator nonetheless. He earned his place in the aviation history books by famously lighting a cigarette while in the air. In an action to be repeated thousands of times over the years, he was then told to go back to his seat by the cabin steward who found him in the plane's toilets.

Now one of the methods of encouraging flying endeavours was to offer prize money. A well-known British newspaper announced they would give £1,000 to the first flyer to make it across the Channel alive. With pound signs lighting up in his eyes, Latham took off on a famous date in 1909 and on the same day achieved a notable first: the first to land on a sea, an engine failure putting him in the drink.* Nonchalantly sitting on his now-floating seacraft, the brave Latham lit up a cigarette. We can only wonder which brand but perhaps it was ... Senior Service. Or ... Capstan. He was picked up by the torpedo boat *Harpoon*. Or was it the harpoon boat *Torpedo*? It doesn't matter. His machine was dragged out of the water and he ordered up another aeroplane for another go, as you do.

On a different famous date in 1909, the gallant figure of Monsieur Blériot strode out to the grass field where history was possibly about to be made. Well, he actually hobbled due to burning his foot on an exhaust pipe.

Blériot had designed his aircraft along what we would see as a traditional format: a fuselage with an engine at the front, two wings, a tail, two tailplanes, a seat for the pilot and a control column for him to waggle about to avoid flying into trees. The bold Louis gazed at his machine. What

* **Aviation term for sea.**

the hell was he thinking attempting this? It was made of rice paper for heaven's sake!

Louis was not known for his chat and often would respond to a question with a shrug. He was French after all. He was laid-back. He had sangfroid and laissez-faire in abundance. But this morning he was especially quiet. Then he seemed to resolve himself to action. As his wife was in England, he kissed his mechanic for luck, then started to quickly run away. Mechanics and onlookers tackled him and marched the bold flyer back to the aircraft before he could try that again.

He clambered into his machine, then stopped what he was doing* and asked a question of his mechanics: 'Where is Dover?' His crew laughed heartily. Then the franc dropped. He wasn't joking. They eagerly pointed away in front, over the coast. Blériot nodded and shrugged. That seemed to be good enough of a steer. And he could always ask someone for directions on the way.

Louis B started the engine and off he went. Up, up he rose into the early morning sky. The plane reached a speed of Very Slow, which was fast enough, and started crossing the grey, cold water below. Blériot's mind filled with excited thoughts. What would he do once in Britain? He'd heard British food was great, so he was looking forward to that. Maybe he'd take in a football match? Or a West End show? Go to Madame Tussauds? The possibilities were endless.

After a bit Louis lost sight of the land he was aiming to land on. He looked down in the cockpit to where it might have been a good idea to fit a compass. Someone had given him a *pair* of compasses but he threw them over the side. Who needs to draw a perfect circle on such a flight? Actually, maybe they would have come in handy if he had to ditch and then fend off a shark. Do you get sharks in the

* **Praying.**

English Channel? He didn't know but he knew he didn't want to find out the hard way.

As he puttered on over the cold, grey void of the deep, deep sea, Louis figured that if he kept going, he'd eventually cross the English coast at some point. It was a big country; how hard would it be to find? This blind faith would serve many well in the future when they hadn't learned how to fit navigational equipment.

Now, Blériot's monoplane had only been flown for about twenty minutes or so before it began its epic flight. So what? You might ask. Well, it meant it hadn't been on a long-distance flight yet. And this might have been something worth finding out beforehand, as when the plane reached halfway it started to overheat. Uh-oh. This wasn't good. But as that famous American philosopher Forrest Gump said, 'Life is like a box of chocolates'. And for the flying Frenchman, these would be Celebrations as suddenly it started to rain, and the precipitating water cooled the hot engine. Was God a Frenchman? It seemed plausible.

Blériot flew on but the wind got up. Louis regretted having the full French breakfast, but it was tasty right enough. He hung on. Soon he saw land. He continued with hopes rising and eventually the gallant Frenchperson soon found himself flying over this land. He touched down near Dover Castle, breaking the undercarriage and the propeller. A crowd rushed to applaud this new hero of the skies. He'd done it! As he received the greetings from well-wishers, a purposeful man strode towards him. Was it the chair of the Dover Planespotters' Group looking for the registration? No. Was it the Dover Castle staff looking for a parking fee? No. This being Britain, it was a customs officer verifying if he'd packed his bag himself and had brought any foodstuffs, explosives or ladies' underwear with him. The last item wasn't *verboten* but just the perverted custom officer's hopes.

Blériot had made history. Fame and fortune was his. The newspaper that offered the prize cleared the front page. The headline was ready:

Channel Crossed by Frenchperson.
How Will This Affect House Prices?

Showing a great spirit of magnanimousness through gritted teeth, Latham sent a telegram of congratulations. And showing the famous French 'never give up' spirit, he got into his machine a few days later and took to the skies. He might not be first, but his flight didn't last. He once again entered *la mer* just short of the English coast. The gallant Latham was not able to enjoy aeronautical success for long. A few years later he was trampled to death by a buffalo he'd shot in Africa. It is not known if he was smoking at the time.

REIMS AIR MEET IS MURDER

In 1909 an event described as the 'Olympics of the Air'* took place. It was the world's first large-scale aerial competition/air show/meet. Officially it was the 'Grande Semaine d'Aviation de la Champagne', and it was a very grand *semaine* indeed. As you might guess, it was held in the centre of European aviation that was France. Over twenty pilots took part, most of them Frenchpersons. The Wright brothers didn't attend, as one of them was washing his hair (strangely, the bald one), but plenty of other intrepid aviators did.

Crowds flocked – who knew Europe had so many undertakers? Five hundred thousand people thronged to see the bold ones take to the skies. The Reims Planespotters' Group

* **By us, just now.**

were there in force, notebooks and wetted pencils at the ready. (Why do people do that? Don't they get lead poisoning?*)

Prizes were on offer, proper ones too, not those cheapy 'gold' plastic ones they hand out at school sports days. Henry 'Henri' Farman flew a machine powered by a Gnome rotary engine, so called because the cylinders and crank-chambers spun round the centre shaft, held by groblets in an anhedral flange-bracket clampset (or something else too technical to understand properly). It was good, and could power a machine effectively, but did mean the craft could tear off in any direction like a startled mongoose, so care had to be taken by the aviator. Farman won the duration competition, trousering the biggest cash prize on offer: 50,000 francs. He was in the air for over three hours, which was spent circling the field, covering a distance of 112 miles. That's what happens when you've forgotten your map. Circling like this was great news though for those wondering what they would do in the future if passenger-carrying craft were too early for their landing slot.

The clean-shaven Hubert 'Smokin' Latham (now dried out from his Channel endeavours) made his play for the top prize – literally, as he entered the altitude competition. Up, up and away he went. The crowd gazed upwards as the aircraft got smaller and smaller. Latham reached an altitude of 509ft. Could humankind survive such heights? They could. Latham returned to the applause of the crowd, none the worse for his epic flight, his return eased by a reward of 10,000 francs.

At one point in the proceedings Eugène 'Genie' Lefebvre flew his biplane towards the crowd. Thoughts rushed into spectators' minds as the plane rushed towards them. Would they feel the chopping motion of the propeller? Would it hurt? These remained unanswered as at the last moment the bold

* **Pencils don't have lead; it's graphite. – Ed.**

aviator swung his plane round and it banked at an angle never seen before. He wasn't crashing; he was demonstrating aerobatics, something that would soon become standard fare for air-show-goers.

The newspaper proprietor Gordon 'Yes, I'm a Real Person' Bennett offered a trophy, the Gordon Bennett Trophy, awarded for the fastest speed over a distance category. This was the final event of the week. American flyer Glenn 'Coe' Curtiss had brought his machine over from America and held it back in readiness. The home favourite, Louis Blériot, had brought a new aeroplane too.

The competition started. They took to the air in their magnificent flying machines and vied for the prize. Who would win? Who would be second? Would enough survive for there to be a third? We would soon find out. After a gripping competition, the winning machine reached the finish line, having achieved the moustache-flattening speed of 47mph. 'Gordon Bennett!' everyone exclaimed when they saw the man giving the cup. GB gave it to the winner: the American, Curtiss. Glenn had done it!

The first aviational event was over. British chancellor David Lloyd George was there and bemoaned how far behind Britain was. (And how there were so few lovely ladies to ~~chat up~~ talk to about the balance of payments.) No one had been killed at the event, but the crowd didn't go home disappointed. They had seen plenty of crashes and Blériot's plane had burst into flames. There would be more chances in the future.

CODY BANKS ON SUCCESS

The British were slow to get going but luckily (like rock'n'roll and online shaming) inspiration came from across the pond. It arrived in the shape of Samuel 'F for Flying' Cody. Cody was flamboyant, brilliant and not actually called Cody.* He had been a Wild West showperson and part of the show involved him shooting at his wife. Luckily for her he was a bad shot. He wore large hats, sported his head hair long and had facial hair fashioned into a style later adopted by modern-day hipsters. The extroverted airperson sported a wide and thin moustache with a long straggly beard. Wax, sir?

When Cody proposed aviation being of use to the military, he was told to go fly a kite, so he did. He showed the potential of person-lifting kites by being lifted above a navy ship by one of them. 'I can see your yacht from here,' he quipped to the naval equivalent of tumbleweed. Samuel F also showed the potential of their lethality by coming down when the ship turned. He 'lost the wind', 'gained the sea' and almost 'drowned'. The former cowpoke was unfazed by setbacks and went on to cross the English Channel, not in the air, but in a boat being pulled by kites. It's a transportation method not widely adopted, sadly.

The army eventually took up his ideas – literally – and Cody made kites for the Royal Engineers. He would have made them for the Royal Flying Corps, but it hadn't been invented yet. Soldiers would be lifted up hundreds of feet into the air in a wicker basket seat to recoin … reckon … recoineu … have a look at the battlefield. Unfortunately, it wielded little useful tactical information as the men had their eyes clamped shut the whole time. You know those theme-park rides that take you up really, really high and then drop you really, really quickly? It was like them, but without the chance of a hotdog afterwards.

* **Interestingly, he adopted the name of that other showperson Buffalo 'Bill' Cody as it was a shorter version of his own birth surname, Cowdery. Maybe at school he was teased as 'Cowardly Cowdery'? It does kind of rhyme.**

Kites were all very well, but they couldn't move about much. A different form of craft was needed and it came in the form of those easily pronounceable semi-rigid dirigibles. Their gas bags were filled with that notorious hydrogen, not the less flammable helium. Helium came with its own problems though: it was always getting used up by the crews doing funny voices. The gas bag was not for the plant-based vegans, being made of 'goldbeaters' skin', which sounds not too bad until you find out it is stretched animal intestines. Basically, these flying craft were very big sausages.

Cody worked with an army colonel called Colonel 'JE' Capper on a dirigible called *Nulli Secundus* (which as all *Harry Potter* fans will know is the spell for 'not being second'). It was the UK's first military dirigible and on a notable date in 1907 it sauntered over London and circled St Paul's Cathedral – anything to avoid paying the admission fee. Londoners stared up agog at this remarkable sight, and packs of dogs sniffed the air, saying to each other, 'Sausages?'

It was a great demonstration of military air power that was only a bit diminished by the wind getting up and overcoming its 50hp-engine-powered progress, at times pushing it backwards. The crew landed at Crystal Palace and the hydrogen was released to deflate the gas bag. Misinformed bystanders rushed forward inhaling, ready to do funny voices, but were disappointed when instead of a laugh they got nausea, headaches and told off for being so bloody silly.

The flight earned Cody and Co. continued funding, and the 'second to none' became 'second to one' as the *Nulli Secundus 2* was built using bits from the *1*. But Cody got bored of airships. He had developed gliders and, realising they could crash at higher speeds if fitted with an engine, wanted to try his luck with powered aircraft. The British Army were vaguely interested, and Cody built the *British Army*

Aeroplane Number One, which was not the most cunningly devised secret code name.

This machine was dubbed the 'Flying Cathedral' because so many prayers were said in it. The prayers initially worked as Cody etched his name in the history books on a memorable date in 1908 when he made a short hop. This wasn't enough. He then flew a distance of 1,380ft, some of it with his eyes open. Luckily, they were open when Cody saw some trees up ahead and he changed course. Into the ground. Turning had slowed the cathedral down and, when a wingtip touched the grass, it was soon followed by the rest of the machine. But he'd done it!

No matter the scrappy ending, this was a truly historic moment. It was the first-ever time a man with long hair who had earned a living as a cowboy had flown a powered heavier-than-air aeronautical craft in (relatively) controlled flight in the UK. The *Gin and Tonic Book of Records* was duly contacted. The military acted quickly, showing the foresight and tactical acumen to be seen at the Somme, Passchendaele and Gallipoli. They cancelled Cody's work, saying airships were the future, not winged craft.

Wily Cody was not dissuaded and built more aircraft. These crashed too but he went on. He was determined to win a contract from the British government, truly believing he was the right man to receive its money. All he needed was the right opportunity.

In 1912 the British War Ministry held a competition to find the best aircraft to be bought for the country's armed forces. Cody's crashed but he built another and won the competition. His success didn't result in long-term prosperity though. The government ordered one of his machines but more of a rival aircraft, the B.E.2, then Cody died in a plane crash in 1913. At his funeral over 100,000 people came to pay their respects to this larger-than-life character – something the great showperson would have revelled in. Even more so if they'd paid a shilling a pop on entry.

OTHER PIONEERS

In those heady days of early flight, there were many pioneers eager to dip their toes into the broken new ground of mixed metaphors and flight. It wouldn't be feasible to ask anyone to research *all* of them, so not everyone is included. Ones we couldn't find space for include Werner 'Herzog' Landmann, who despite his name refused to land, setting a record for an endurance flight of over twenty-one hours. That's almost a day! Then there was Reinhold 'Dive' Böhm, who was first to fly for more than twenty-four hours – in one go – and who broke the 100m sprint record on his way to the lavatory after landing. Also, there was Roger 'Over and Out' Sommer, who had a cracking beard that made him look as though he was hiding behind a merkin,* and who set records for the number of passengers he took aloft.

The ones we did find time to find are:

OTTO 'FLIEGER' LILIENTHAL

When it comes to pioneers who created successful gliders, no name is etched higher in the *Liebfraumilch Book of Records* than Otto 'Glider' Lilienthal. This German thought that in order to fly, more must be understood about birds. To this end he lived up a tree and ate nothing but worms. With this knowledge, and a newly acquired wish to fly over noisy neighbours' charabancs, 'Birdy' Lilienthal felt the next stage

* **Ask your medieval ancestors.**

was to build gliders. By flying them he would therefore expand his knowledge, and this he did, flying more than 2,000 times in his creations of bamboo and cotton. His technique was simple: find a hill and jump off it. In the air Otto would hang suspended and control the glider's movement by shifting his body, a bit like when you have an itchy behind while stuck in a team meeting. Like many of the early aviation pioneers, the intrepid aviator would not end his days an old person, propped up in bed, patting the heads of his grandchildren, regaling them with tales of a long and productive life. One of his gliders was a biplane and, when the top wing came off mid-flight, he fell to the ground, dying the next day.

PERCY 'NO, NOT PILCHARD' PILCHER

The Bat, the Beetle, the Gull and the Hawk might sound like the title of a children's picture book but it's not. Yet. They were gliders built by Percy Pilcher in the late 1890s. Although sounding like a footballer for Accrington Stanley, Pilcher was an intrepid pioneer in the field of unpowered aircraft. He was inspired by Herr Lilienthal and could regularly be seen floating over Kelvingrove Park in Glasgow in his various designs. Sadly for the bold Percy, the number of successful landings didn't match successful take-offs. He died following a test-flight crash.

OCTAVE 'KING' CHANUTE

Octave was a civil engineer (the ones who stand up when a lady enters a room and always say please and thank you). Chanute was too old and sensible to venture into the air himself but enabled others to do so. He developed a way of bracing the wings so they would not fall off in flight (high up the priorities among pilots). He also worked on adding extra wings so that, if one set broke off, there would be a back-up. He wasn't daft was our Octave and he helped so many others he was described as 'the father of aviation' in America. But not in Britain, where we had our own father of aviation, thank you very much. Or France for that matter.

ALBERTO 'SANTOS' DUMONT

Keen-eyed readers will remember his name from earlier in the book when he was flying ballons. After his fun with the inflated floaty things, the great Alberto had a look at gliders. He thought, 'Nah, they look dangerous,' but in the fun way of the times, he thought, 'Ach you only live once.' He test flew a Voisin glider – or rather test sailed, as his flight down the Seine ended up *in* the Seine. He was not deterred and flew a larger, drier machine.

It was thought at the time (1906) that no one had flown an aeroplane, as news of the Wright brothers' first flight hadn't been believed by many in Europe. One Wright-denier was Ernest 'Holy' Archdeacon, a pioneer who put up a prize for the first flight over 25m. Dumont fancied trousering the prize (which was a silver trophy so we can assume he had big trousers). On a famous date that would echo down the aeronautical history decades, the resplendently moustached Alberto became the first person to fly a personned and powered aircraft in sustained flight in Europe. (We realise we probably should have mentioned this achievement earlier but do you know how much copy editors cost?* Their not cheape.)

* **No idea.**

Alberto's 14-bis machine was a strange-looking craft that appeared to be flying backwards, as the box-kitey wings were positioned at the back, alongside the engine, while out the front at the end of a long fuselage was a box-shaped area. The pilot stood up. Not because a lady had entered the airfield, but because there was no seat. If it sounds unusual, it was. But it was enough to keep him aloft for the required distance to win the Coupe d'Aviation Ernest Archdeacon. He'd done it!

SAMUEL 'L' LANGLEY

The American Langley was at the Smithsonian Institution when he received an enquiry from a pair of brothers from Ohio who wished to be given the 'Wright' advice. Yes, it was them, the Wright brothers! They had written asking for books on aviation research, but fancying building his own flying machine, Langley sent them books on basket weaving, astrology and stamp collecting. They wrote back saying, 'Ha ha good one', so he relented and did the 'Wright' thing, and the intrepid brothers got their books. With all the drawings scribbled over.

Langley was no slouch in the aviation pioneering business, although he risked his chances by calling his flying machines 'aerodromes'. When he told people he was building an aerodrome that would fly, they laughed in his face. 'What an idiot that Langley guy is,' they'd say to each other, and if he overheard, he'd be a bit hurt, but he knew he'd be proved right one day and would have the last laugh. His early aerodromes were unmanned, and they could – pun alert – perhaps have been called '*aerodrones*'.

Langley's aerodrome/drone was duly prepared for flight and on a date in 1903 it was powered up on its launching location: a houseboat on the Potomac River. Langley tested his machines over water as he just loved the sound of water lapping against the boat's hull. It can be relaxing right enough.

A catapult was installed to ping the ~~doomed~~ hopeful contraption along rails, to ensure a short take-off run. Sitting ready to become the first person to fly in a powered machine was Langley's assistant: Charles 'Very' Manly. When the catapult was triggered, Langley's machine hurtled forward, then down, right down into the water, allowing Manly a chance to practise his front crawl.

Undeterred, another attempt was made a few months later. The catapult was triggered. The machine hurtled off again. Onlookers raised their eyes to the sky to follow its progress. They were looking in the wrong place. It was in the water. Manly was trapped under the wreckage and in danger of suffering from hypothermia. Onlooking funeral directors edged nearer but a nip of something alcoholic revived the intrepid aviator. Langley insisted that the launch machinery was at fault but people couldn't hear him for the sound of their own laughter. He wanted to continue development but his requests for more money were met with a simple response: no.

BARONESS 'RAYMONDE' DE LAROCHE

Despite the gender-determinative-sounding name, the Baroness most definitely identified as she/her and is thought to be the very first woman person to pilot an aircraft. Laroche was also the first woman to get a French pilot's licence. She flew at Reims but, as women were considered too delicate* to take part in combat flying, was denied a

* **By men who spent most of their time on the ground.**

seat over the Western Front in The Big One. She continued flying but this came to an abrupt end when she died in an aircraft crash in 1919.

GUSTAV 'WILHELM' HAMEL

As you would suspect, Hamel was from Japan.* He made the first trial of an aerial postal service. It was between Hendon in London and Windsor outside of London. Hamel aimed to deliver a letter fit for a king: King George the Fifth, who lived in Windsor at the time. On the chosen day it was windy, and Hamel's friends urged him not to fly.

'The king must get his mail,' Hamel responded.

'You do know we have cars and horses and the like?' they reminded him. But there was no stopping the German. He took off in a Blériot and nineteen minutes later landed in the same one in Windsor. It was successful in that the pilot arrived in one piece. No dogs could get near *his* trouser legs. Hamel was flying from France to Britain in 1914 when he disappeared. A couple of months later, fisherpersons in the Channel found a pair of goggles on the sea. They were Hamel's. He had died.

EUGÈNE 'GENIE' LEFEBVRE

The neatly waxed moustached test pilot for the Wright brothers in France ended his days while flying one of the Bros' biplanes in 1909. His place in the *Côtes du Rhône Book of Records* was sadly etched as the first-ever pilot to die in a flying accident.

* **<sighs> Germany. – Ed.**

ADOLPHE 'JOHNNY' PÉGOUD

Crowds wanted more and more thrills from their bold aeronauts. They'd seen them fly around in circles and then crash. They wanted to see them do aerial stunts, and then crash. One who answered the call was the twirly-moustached Adolphe Pégoud. He would go up into the air to 'loop-the-loop'. This involved the pilot flying upwards in a big arc until he or she went upside down, then carrying on downwards until he or she was back on the level where they started – minus all the change they had in their pockets. Adolphe also did spins and tail slides, manoeuvres that would give the stoutest heart a little flutter as they are essentially CRAZY! Another trick Pégoud performed was to jump out of an aircraft. You'd think this would be a one-off trick, but he avoided the certain death by not only wearing a parachute but using it.

His derring-doings saw him trouser plenty *d'argent* as pilots could enjoy good money from aerobatic performances, providing it came in cash on the day and not as a life insurance cheque. Pégoud passed on his knowledge and instructed young novice pilots, but while it is rewarding seeing your charges taking off, it isn't rewarding seeing them shooting at you, which is what happened to Adolphe. In the First World War Pégoud was good at shooting down aircraft and became the very first 'ace', i.e. a pilot who is dead good at flying. Sadly for him and the Allies, Pégoud was shot down and killed in 1915 by one of his students. That's gratitude for you. And the cruel reality of war.

ALFRED DE PISCHOF

The bearded Pischof is included not only because he was an early pioneer of powered flight and flew his own biplane but because his surname sounds like something you'd say to someone who told a doubtful story.

THE NIEUPORTS

Another bunch of flying brothers were the Nieuports: Édouard (waxed moustache) and Charles (full, bushy). They designed, made and flew aircraft with an emphasis on streamlining, to make them look nicer (and go faster). Fuselages were enclosed in material, and wheels were made as small as possible. Sadly, this didn't prevent them crashing, with Édouard in 1911 and Charles two years later piling in. Their name survived as a great aircraft manufacturer, making some of the finest fighters in the First Great World War.

ARMAND 'IN HAVANA' DEPERDUSSIN

It is a dangerous game listing the 'best-looking' aircraft, as aircraft beauty is in the eye of the beholder. It's a subjective thing but you'd have to have your eyes painted on to say the 1913 Deperdussin Monocoque was not beautiful. Where aircraft were generally biplanes, this was a monoplane. Where internal struts formed the fuselage, this used a hollow shell similar to what you would see nowadays if you sawed through an aircraft (as long as you had asked permission from an adult first). Where pilots used levers to control the aircraft's control surfaces, this used a steering wheel. As two fictional pilots would later say in an unrealistic movie about navy fighter pilots, this little beauty had a 'need for speed'. It was the fastest thing in the sky, setting several records, at 108mph and then 19mph faster. It might not seem a lot to the speedsters of today but you can do a lot of damage flying into a tree at 127mph - and not just to the tree.

The company that built it was owned by the silk-importing entrepreneur Armand Deperdussin (full). Sadly, Armand did not live a long life and wasn't able to enjoy the fruits of his aeronautical labours, labours that had seen his fast machines winning Gordon Bennett's and the first Schneider Trophy. Armand got into bother with money - he didn't have as much as he said he had - and wanted to get some by non-legal means. He was found guilty of fraud and died in 1924.

GLENN 'COE' CURTISS

Many of the early pioneers caught the flying bug early. Literally, in the case of Glenn Curtiss, who built an airplane called the June Bug. Curtiss had all the qualities needed for a bold aviator pioneer. He was technically gifted; he was courageous; but, more importantly, he had a fine moustache. He also had an aircraft that could fly.

His June Bug won the first prize offered to an American aviator, the Scientific American Cup, which, unusually for an American competition, didn't have the prefix 'World' ... In front of an expectant crowd on that date so dear to Americans: 16 June* 1908, Curtiss flew a distance of over 1km. He had won the Cup! And 25,000 dollars, which back then was a lot of money. Curtiss went on to fly his airplanes over

*** <sighs> It was 4 July. - Ed.**

water, on it and also on the decks of navy warships, ushering in a future that would include aircraft carriers.

ALEXANDER 'GRAHAM' BELL

Bell was famous for inventing the phrase 'Give me a bell' – meaning 'give me a telephone call on the telephonic invention devised and patented by me: Scotsperson Alexander Graham Bell'. Alex was also interested in aviation and formed the AEA (American Experimental Acronym) in order to promote aviational advancement. He employed a certain Glenn Curtiss, whose June Bug was an AEA product. That's very interesting, isn't it? The Wright brothers weren't happy at what they saw as patent infringement and a lucrative legal battle began in which the lawyers won.

Bell's AEA made the first practical aileron, which meant years of trying to say it without sounding plastered. Aileron. Aileron. Try it. Aileron. You're my bezzie mate you are.

Although it looks the same, the 'Bell' of Alexander Graham 'Bell' is not the same 'Bell' in the 'Bell' Aircraft Corporation. That was a different 'Bell'. The 'Bell' Aircraft Corporation was formed by Larry 'Bell', who had worked with Glenn 'Not Curtiss' Martin in the Martin aircraft company. Not the Martin-Baker company. That was a different one.

LINCOLN 'CONVERTIBLE' BEACHEY

The unmoustached Beachey was hired by the ubiquitous Glenn Curtiss to demonstrate his aircraft. Lincoln demonstrated their ability to crash by doing that three times while learning to fly them. Not a great advertisement, it has to be said.

Beachey persevered and became a famous show-off. He could loop-the-loop, recover from spins, set altitude records and wow the crowds by flying over the Niagara Falls. Some thought it would be even more impressive to fly *under* them. Everyone's a critic. Beachey raced trains and, apart from a switch to a bus replacement service halfway through, easily won. He was some boy, was our Lincoln.

But there was a dark side. Beachey was wracked with guilt over deaths he felt responsible for. Several pilots had tried to emulate his 'dip of death' dive but only proved the literalness of its description. One time Beachey's plane swept several onlookers off a shed roof, and one was killed. Did it put punters off coming to see him? Not a chance.

Beachey retired from flying three times, but each time had returned, saying, 'Just when I thought I was free, they pulled me back.' On a sad date in 1915 he retired permanently when his brand-new monoplane broke up in flight off San Francisco Bay and the bold Beachey hit the water. He survived the crash but drowned.

GRIFFITH 'ALE' BREWER

Although not as well known as some of the others, Brewer has his place in the *Warm Beer Book of Records*, being the very first person from Britain to fly in a powered aircraft, not as a pilot but as a passenger, being flown by one of the Wright brothers in 1908. Brewer was great friends with the Wrights and assisted them in their battle with the Smithsonian Museum, which had presented Langley's damp aerodrome/drone as being the first machine to fly. When it hadn't.

JORGE 'PLUCKY' CHAVEZ

A prize was offered for the first aerial crossing of the Alps, from Switzerland to Italy. Plucky Peruvian pilot Jorge 'I'm Plucky, Me' Chavez took up the challenge in 1910. In a Blériot XI, he took off from the country of cuckoo clocks and hard-to-eat mountain-shaped chocolate. As onlookers watched, Jorge headed for the Simplon Pass. Would he fly into the rock formations, a downdraft dashing his wood-and-fabric machine into pieces onto the aeons-old stone walls? The onlookers crossed their breath and held their fingers. Chavez's machine changed its attitude in the sky. It rose and rose and suddenly ... made it through. This gutsy pilot lowered the nose and started to descend towards Italy. His descent was too rapid though and he crashed, dying of his injuries several days later. Chavez is remembered by having an airport named after him. So that's something, although not much use to him on a personal basis.

AV 'COD' ROE

Despite being clean-shaven, Roe was a very successful aviation pioneer. He began his aviation career by winning £75 in a newspaper aviation competition. It was a model-building contest and Roe managed to not stick any pieces to his fingers with the glue. His completed model flew a hundred feet.

With the prize money he then built a bigger aircraft, one that a human (him) could fit in. Roe would run up the engine of his biplane, then when reaching full power it would be released to trundle along and make 'hops'. This interested the beer-loving aviation enthusiasts: a beer-making aircraft? A dream come true! But no, it was not producing ale. It was making little jumps into the air, followed by the inevitable crash.

At one time AV encountered opposition from the local authority, who thought his aerial experiments were a public nuisance. Roe got round this by ... getting up early. It doesn't always need super cunning to be ahead of the game. Keen and enthusiastic, he once cycled to France to meet with American aviation pioneer Mr Wright. Roe arrived wet and tired. The American said to him, 'Why didn't you fly?' Roe quietly clenched and unclenched his fists. But the meeting gave the Briton great insights: this Wright had achieved success without a moustache and, by golly, he would too.

On his return, Roe kept on pursuing his dream. It wasn't easy. He was skint and had to cover his aircraft's fuselage in brown paper, the same way a child covers their school books. However, on a famous date in 1909 that would echo etc., he made the first flight by a British person in Britain in a British-made aircraft. And it didn't crash! In a clever marketing ruse, he'd called his aircraft the Avroplane. It was a triplane, which was actually one plane but with three wings, named from the Latin 'triplane' meaning one plane with three wings.

Roe formed a company using that common asset of the early aviation pioneer – his brother. Avro went on to make great aircraft such as the Avro 504 and the Lancaster. They also made the Manchester. Many aviation pioneers were immortalised by having streets named after them: Cunningham Way or Cody Avenue but AV Roe never had a Roe Row or a Roe Road. However, he did get a popular children's song written in his honour about going gently up a stream.

JOHN 'THEODORE CUTHBERT' MOORE-BRABAZON

JTCM-B earned his place in the annals of aviation history (and this book) by being the first British person in Britain to fly a powered aerocraft in Britain. Now you might think, 'Wait a minute, Writer of This Book, didn't you just say up there ↑ that AV Roe did this?' Ah, but you see, JTCM-B flew a French-built Voisin as around this time the French were making many of the aircraft being flown. They were also teaching pilots how to fly. This was not universally welcomed, with complaints about them flying on the right-hand side. It caused chaos at roundabouts.

JTCM-B then got his pilot's licence – the very first person to receive one in Britain. Well done, sir! But wait, didn't he just fly? It's a bit like Nigel Mansell getting his driver's licence after winning the Monaco Grand Prix. If his Williams flew and did 25mph.

A newspaper offered a prize for the first British person to fly a British circular mile in Britain in an all-British-built aircraft. There was only going to be one person to do this and JTCM-B's circular flight in a Short biplane saw him trouser the £1,000 prize. He would later win another grand from the paper for a long-endurance flight of 17 miles.

The bold Johnny famously flew a pig in a basket, adorned with a sign saying 'I am a pig in a basket'. The reason he took up a porcine passenger was that, before his circular mile flight, he overheard someone say he had as much chance of achieving it as a pig does of flying. Despite 'bringing home the bacon',* he didn't get £1,000 for doing it.

JTCM-B managed to avoid death by aerial accident and was made Lord Brabazon of Tara, being a big fan of *Gone with the Wind*. He must have upset someone as Lord B was

* **You're welcome.**

given the thankless job of supervising British civil aircraft production after the Second World War.

CS 'SWISS' ROLLS

Rolls was a keen balloonist, although his name would live on in another form of transport: cars. Along with a person called Royce, he formed Aston Martin, or was it Bentley? We should look that up. In 1910 he entered the aviation pantheon when he was the first to make a return flight over the English Channel. You really should check if you've left the oven on before departing. The moustached aviator re-entered that pantheon later that year in a non-good way when he became the first UK aeronaut to be killed in an aircraft crash. He was flying at Bournemouth when the rudder came off, which is never a good thing.

GEOFFREY 'DE' HAVILLAND

Geoffrey designed and built his own aircraft before being recruited by the Balloon Factory. He went to work at the Airco company and received a royalty for every aircraft built to his designs. This was acute business acumen and saw him trouser a decent wedge. Geoff then set up his own company, which caused problems for those not sure how to spell it. Was the 'de' in lower case or Initial Caps? Despite this, de Havilland went on to make classic British aircraft: the Tiger Moth, the Mosquito and the Comet. Well done, Geoff. Interestingly, to continue the *Gone With the Wind* theme,* Geoff's cousin Olivia de Havilland acted in this film.

* **Is there one? - Ed.**

SHORT BROTHERS

Yes, another brother combo: Horace (no moustache), Eustace (moustache) and Oswald (no moustache). They had begun building balloons but in 1908 set up Britain's first-ever aircraft manufacturing factory off Kent on the Isle of Sheppey. Eustace Short had obtained a licence to build Wright Flyers. They made six, which sold like petrol in a panic-buying panic. Short Brothers went on to to make seaplanes and flying boats, which were aircraft that could take off and land on water, a useful function for operators not wishing to pay airfield landing fees.

In a strange twist on the normal outcome, Eustace died in an aircraft, but not in a crash. One day he landed his Mussel seaplane and taxied in but when he didn't get out, or switch off the engine, onlookers thought something was up. Sadly it was Eustace going towards the Pearly Gates as, on reaching the aircraft, they discovered he had died with his flying boots on.

HARRIET 'FRED' QUIMBY

Quimby was the first woman in America to get a pilot's licence. She wasn't finished, as the bold Harriet then became the first woman to fly herself over the English Channel when in 1912 she flew herself over the Channel. This achievement was overshadowed by the sinking of the *Titanic* a few days before. Typical, isn't it? You do something great and then a ship carrying thousands hits an iceberg and plummets to a watery grave. Huh. Sadly, Harriet herself was to follow suit as, a few months after her epic flight, she was flying over Boston's Dorchester Bay when she and another pilot fell out of their seats – seat belts being something of the future – and fell into the water.

HENRI 'WILL GO' FARMAN

The moustached and bearded Farman was a bit British and a bit French. People constantly got his name wrong and called him Henry Farmer. This riled the aviator but he acted unruffled. It became so common that he thought for a while of changing career and getting into agriculture. (He had spent a lot of time in fields due to landing short, right enough.) One field he flew around was the airfield at Issy-les-Moulineaux, which is in France. His 1km circuit in a Voisin biplane in 1908 won him 50,000 francs as winner of the Deutsch-Archdeacon Grand Prix, which at the end of the day is a nice little earner.

Farman also etched his name into the *Châteauneuf-du-Pape Book of Records* by becoming the very first passenger in a powered aircraft. In that same year of 1908 he was taken aloft by Léon 'the Lion' Delagrange (neatly trimmed) in a Voisin. As if these achievements were not enough, a further chapter in the book was etched later that year when the bold Henri was the first pilot in Europe to make a cross-country flight.

Something we just found out was that Farman had two brothers and they formed the Farman aircraft manufacturing company, which made aircraft for the First Big War. Two of their designs were the Longhorn and Shorthorn biplanes, so named because they liked to spend time munching grass. No! Of course not. It was because they tasted real good on a barbecue. No! It was because of the wooden skids that ran out the front of their undercarriage.

All three Farman brothers lived into their seventies and a couple went further, into their eighties, but no further.

TOM 'TOM' SOPWITH

TOM Sopwith wasn't shouting his name, these were his initials, although he did shout a lot as his hearing was affected by being close to noisy engines all day long and proper PPE was a thing of the future.

In 1910 Sopwith entered a competition to win £4,000 for the person who could say in fifteen words or less why cornflakes were the best cereal. Nah, it was a flying one, you sillies. A baron offered the cash prize for the longest flight from Britain over a continent. Luckily for the competitors the continent was Europe. The competition was to encourage British aviators as the country wasn't leading the way as it damn well should, sirrah.

Tommy had faith in his abilities and a flask filled with a meat-based drink. As he lined up for take-off he was aware of being watched. He was glad of the support from the public, but it was only when the bold aviator felt someone licking his neck did he cotton on to the fact that his adoring public was really a labrador keen to tuck into that flavoursome beef drink. With a 'Get down, Shep,' TOMmy shooed the mutt away and got on with his flight. Despite his compass not working, he flew 169 miles into the heart of Europe. After landing he went up to a local: 'IS THIS THIRLMONT IN BELGIUM? IF SO, I'VE WON FOUR GRAND.' It was, and he had.

CECIL 'MR' GRACE

One of TOMmy's competitors was Cecil Grace. Unusually, he wasn't a brother; it was just Cecil. Cecil made it over the Channel but there was bad weather and he chose to fly back and try another day. There were not to be any other days, as he disappeared in fog and was never seen alive again. If the sea is a cruel mistress, then the sky, combined with the sea, is an absolute &*$#@rd.

MANCHESTER OR BUST

Probably should have mentioned this much earlier but better late than never.* In 1906 a competition offered by a newspaper offered £10,000 to the first aviator to fly from London to Manchester. Pilots rubbed their chins: if this was for Manchester, how much would they get to go to Scunthorpe?

The brave pilots had to cover the 185 miles in twenty-four hours. They could only land twice for petrol and Ginsters, and the aircraft had to land within 5 miles of the destination.

With aircraft not being that reliable or safe, the newspaper's money was safe until, in 1910, two competitors had a go in what was the world's first-ever cross-country aerial race.

The Briton Claude 'Grahame' White (no moustache but he did wear a bow tie) took off in a Farman biplane. He flew 115 miles and then suffered an accident that dealt a blow to his hopes of pocketing the cash. Unusually, his plane didn't crash while in the air but while parked on the ground, being damaged by high winds. Hangars weren't really a thing back then.

The bold Claude wasn't put off. He tried again. This time he had a direct competitor. Louis 'Competitor' Paulhan was like many other pioneers of the early twentieth century, in being French and moustached. He became known as the 'King of the Air' when he set an altitude record of 4,165ft in America. Would he become king of the London-to-Manchester air route? Time would tell.

* **Strangely never adopted as a strapline by any low-budget airline.**

The two agreed to a gentleperson's agreement. They would tell each other when they were about to set off, in order to make a competition of it. However, Claude was resting in a hotel when he heard that Paulhan had taken off. Paulhan later said he had instructed somebody to tell somebody to inform so-and-so and they didn't but they told their mate Pierre, who thought he *wasn't* to tell and so it just didn't get done. Claude raced to his machine. He was already an hour behind. Louis and Claude raced through the skies at speeds approaching 40mph, determined to gain fame and, more importantly, the cheque. The two gallant gentlepersons flew on through the darkening evening until both landed for the night.

Claude then did something that would echo down the aviational decades: he decided to fly at *night*. He reasoned this would solve any fear-of-heights issues. If you can't see the ground, how do you know how high you are? His back-up team looked at each other. Some blew through their cheeks. It was his little pink body that would be connecting with any unseen solid ground, so it was up to him.

Claude's team of willing helpers came up with an idea to guide him on his way. In a steam-driven car (this is actually true), they drove ahead with their car's headlights showing the way. He could then look down and see them. It's a clever idea that relies on planes being slower than cars and therefore is not much seen these days, although some of the private hire taxis round our area could give anyone a run. But we are getting diverted.

The two intrepid pilots' endeavours captured the imagination, with great attention from the public on when they would crash. No! When they would reach the finish line. Punters would wave and cheer as the flimsy craft passed overhead. Reporters would scribble notes, checking they had the little card with 'Press' on it tucked into their hatbands. Undertakers crossed their fingers. Surely today would bring cheer from the heavens. There had never been excitement like this.

Disaster almost struck when Claude accidently hit the ignition switch with his coat and turned the engine off. Doh! But he recovered and intrepidly flew on, trying to see those damned car lights. But his courageous attempt ran into trouble: up ahead lay high ground and his aircraft was underpowered.

- Who would win – the gallant Frenchperson or the gallant Briton?

- Who would be the victor – the one who had sneakily taken off first or the brave and upstanding fine citizen who hadn't?

- Who was going home disappointed from *Strictly Come Flying*?

Meanwhile, on the ground, Paulhan readied his aeroplane for flight and took off just after four in the morning. Claude was still behind him. A strong headwind and a stuttering engine had slowed him down and then took him down. He landed short of Paulhan's position. Would he have time to repair it and get back in the air? Paulhan took off with his rival behind him. So, the answer to that was ... no.

The nation awaited, open-mouthed and agog. Those gathered at Manchester saw a small speck. Then they cleaned their glasses – and saw another speck. This turned out to be a small aircraft approaching. It got close, and bigger, until it got to full size.

Who had won?*

* **WHO WAS IT!? – Ed.**

It was the Frenchperson. Paulhan had done it! He arrived twelve hours after leaving London, in a time looked on with envy by current M6 motorists. Claude stood waving his clenched fist at the skies, the holder of his now-broken dreams.

Later that year the daring duo competed again to fly the greatest aggregate cross-country flight.* Would Claude gain revenge? No. He was second, losing the £1,000 to the Frenchperson whose guts he really must have hated by now. Claude had lost the race, but he earned his place in the hall of fame. They made a waxwork of him in Madame Tussauds. And that's what really matters.

MORE RACES

Races were definitely the new-fangled thing of the day and there were plenty around.

PARIS TO MADRID

This 1911 race didn't get off to the best start, as pilot Louis 'Émile' Train's engine started playing up and so he prudently turned back to the field. Fearing he'd crash into some soldiers who appeared in front, he pulled up. And crashed into a party of onlookers who included the French war minister, who died. *Merde!* Another fifty people were injured. *Très merde.*

The race wasn't cancelled as things were different then. One of the competitors (Eugene 'Competitor' Gilbert) was flying over the Pyrenees when he was joined in formation by an eagle. Gilbert wasn't eagleophobic but he didn't like the

* **Whatever the hell that is.**

look in the bird's eye. Just then he remembered he had a gun. His team had provided him with it in case he landed in the Wild West in the 1870s. He blammed away at the surprised bird of prey. The eagle, thinking there were easier ways to get something to eat, dived off towards a field of cute but tasty little sheep. Not affected by hungry eagles, trimly moustached Frenchperson Jules 'Winner' Védrines went on to win this race by being the only one to finish. A win's a win.

PARIS TO ROME THEN TURIN

Next up was a jaunt to Turin via places such as Rome. Unfortunately, bad weather prevented the flyers from reaching the capital city of Piedmont, where automobile companies such as Alfa Romeo, Fiat and Lancia have their headquarters as well as famous football team Juventus and Torino, and where the mythical Turin Shroud is held. The race was won by Frenchperson André 'Winner' Beaumont (who was actually a French naval officer called Jean Conneau but didn't want the taxperson to know about his side hustle earnings) and no eagles were shot at.

The trimly moustached Frenchperson Roland 'Tennis' Garros took part in the race. He didn't win but did reach Rome, which was a darned sight more than most others. Garros went on to be the first to fly over the Mediterranean. No reason given but everyone needs a hobby. His aviational exploits resulted in a tennis court being named in his honour. That's got to be better than a waxwork statue. Sorry, Louis Paulhan.

CIRCUIT OF EUROPE

This race took intrepid money-chasers on a 1,000-mile route. It was touted as a circuit of Europe, but this was a bit hypey. It really only included four countries: France to Belgium to the Netherlands to France to Britain and back to France. It was really a Circuit of Some of Northern Europe.

While some spectators had been injured at the start of Paris–Madrid, it tended to be the pilots who came out worse. At this race three pilots died, including one whose plane caught fire in front of the crowd, who then had to sit and watch him expire. Now we've all seen Piers Morgan on breakfast TV but that is really grim watching. Many were attracted to aviation as a leisure pursuit because of the thrill of being close to danger but there's a limit. It can really put a dampener on anyone's day out.

Louis 'Émile' Train, who had caused the carnage at the Paris-to-Madrid race, suffered his own misfortune, although it wasn't anything as bad. He was heading to London and got lost. He landed to get directions, but without brakes the plane rolled backwards down a field and into a fence. His machine was so damaged he had to abandon, unlike that man Beaumont/Conneau, who won.

CIRCUIT OF BRITAIN

Britain wasn't going to miss out on the fun. The race to Manchester was followed by the Circuit of Britain, offering a whopping £10,000 prize to the winner. Competitors would fly from London to Edinburgh, then brave a potential forced landing in West Lothian on the way to Glasgow. If still alive and aloft, they'd head down south to Bristol, then back along to London. The 1,000-mile route had to be completed inside twenty-four hours of flight time. This required an average speed of taking the 1,000, then taking the twenty-four and looking at it and wondering why nothing you learned in school has stuck.

Anyway, thirty pilots entered and on the day of the race the gallant flyers took off from Brooklands. Seventeen completed the first leg to Hendon. The next day a carpet of fog greeted the aviators. This was a boon for the more nervous flyers as it meant they could remain unfound by their ground crews. The race became a two-horse contest between the perennial Frenchpersons Jules Védrines and André Beaumont /Conneau. They swapped positions on the leader board until finally Beaumont/Conneau reached the finish line ahead. This made him the winner.

Védrines etched his own place in the *Pernod Book of Records* by becoming the first pilot to go faster than 100mph – and not in a death dive. He did die in a plane crash though, in 1919, after surviving the First Terrible War.

AIR SHOWS IN THE UK

Air shows are now a regular feature in aviation enthusiast calendars. Few can resist the smell of jet fuel and lure of the pre-printed car-park reservation voucher. But they weren't always around. If you went back to 1907, you would not find any to attend. If you went back to 1908 – nothing. If you went back to 1909, then you'd be spoiled for choice as two venues – Blackpool and Doncaster – vied to hold the first British aerial competition/air show/meet. Blackpool was a popular seaside resort on the north-west coast of England,* while Doncaster was a future city in Yorkshire. Blackpool's

* **It still is. – Ed.**

event was the official one, sanctioned by the Royal Aero Club's Official Air Show Sanctioning Committee.

~~Farmer~~ Farman and Latham took part. At first they just sat around with the punters as the weather was very bad. It was windy and persisted down for several days. At times the crowd numbered one. At one point someone rang up to ask what time the flying was due to start. At which the official answered, 'When can you be here?'*

But the intrepid Latham was determined to put on a show for the soaked and bored onlookers. He climbed into his Antoinette monoplane and prepared to take off in what was described by an observer as a 'gale', and who are we to doubt this? As he ran up the engine the plucky pilot was surprised as anyone when a gust of wind lifted him and his aeroplane up into the air without any sort of a take-off run. As the monoplane went up, the crowd stood or sat agog and open-mouthed. Would they see Latham returned to Earth in a heap of broken and twisted wreckage? No one had flown in such conditions before, so there was a high chance. Wait, wasn't he going backwards? The wind was so strong it was blowing the aviator in the reverse direction to normal flight. Hopes of seeing something tasty were dashed as the courageous airman landed safely.

At Doncaster the great showperson Cody was in attendance. Being a cowboy, he was intrigued by the prospect of visiting what was described to him as a one-horse town. Cody was so taken with the Yorkshire future city, he elected to become a British citizen on the spot. The Union Jack was raised, *God Save the King* was played and mutterings were heard about the awful weather we've been having.

The stand-out moment at the show was when the bearded Frenchperson Huber 'Le' Blon narrowly avoided ploughing

* **Long-time resident, Old Jokes Home.**

into the crowd. A gust of wind* sent him directly towards a clump of spectators. A quick decision by the former racing car driver to 'give it some welly' (but in French) resulted in the machine lifting up and over the crowd. He still crashed but didn't cause any death or carnage.

The Blackpool crowd were gutted when they heard of this. They were also gutted when they heard Cody had also piled in, being thrown clear of his plane like an unseated young cowpoke flung from a bitter 'n' twisted ole' mustang out on the Texas prairie.

In August a year later, an aerial competition/air show/ meet was held at Lanark in Scotland. Two hundred thousand people came to see That Man Cody, among others. A couple of the aircraft being transported from the previously held event at Blackpool went missing in transit. There is no evidence to suggest that traders at the world-famous Glasgow Barras market were flogging these machines alongside tea towels, sport socks and totally official copies of Gilbert and Sullivan's latest releases. None whatsoever.

Oh, for those hanging on to find out which city had been first to hold an air show in the UK. It was Doncaster. So, well done, Doncaster!

RETURN DESCENDER

Before we go on any further, we should mention getting out. A good way of getting out of a failing flying machine is by parachute. As we'll remember, André Jacques 'Me Again' Garnerin and others not in his family had made successful

* **Meghan Trainor might have sang in 2014 about it being all about the bass, but in the early days of flying it was all about the wind.**

descents under a parachute since the eighteenth century. The early parachutes weren't as portable as the ones we're used to seeing in internet videos of parachutists wrapping themselves around telegraph poles. They were a bit more cumbersome.

In 1910 an Austrian tailor called Franz 'Tailor' Reichelt (full handlebar) tried on a parachute suit. 'So what?' you might say. People try on clothes all the time. But Franz's suit was to have more of an impact than a new set of threads for a night out, for he had made himself a home-made ensemble of silk and rubber that he reasoned would provide enough lift to allow a safe descent from a height. He jumped from a height of 10ft and survived. He jumped from 26ft and broke a leg. Emboldened by his 'success', he figured a bigger altitude was required to fully show the potential of his invention. Living in Paris, there was only one place to ~~kill himself~~ demonstrate to all and sundry how good his parachute suit would be.

On an infamous date in 1912 the bold tailor climbed to the first floor of the Eiffel Tower. He climbed up onto the railing and held his position. He took his time, surveying the ground below. He looked more like a child wearing his dad's overcoat than a person about to provide proof of concept of a new safety device. ~~Future witnesses to the inquest~~ Bystanders tried to dissuade him, saying there's loads of other ways to die senselessly. Had he thought about making his own plane? Joining the Foreign Legion? Getting a crocodile as a pet?

But Franz's mind was set. He jumped. His descent was filmed but they used up less film than Franz would have liked, as his downwards trajectory was the same as if he'd shoved a bag of lead bars over the railing. Film showed the imprint his body made in the Parisian soil. He had deid.

A month later a successful jump from an aeroplane was made in America. Captain 'Albert' Berry was at Jefferson Barracks, Missouri, or rather 1,500ft above it in a plane piloted by Tony 'Surname Doesn't Rhyme With Anything' Jannus when he jumped over the side. Berry's parachute was attached to the front of the plane and fortunately it came with him. He tumbled for about 500ft before the parachute opened properly. Berry landed on the ground not long after. His name was duly etched in the *Budweiser Book of Records*. When asked if he would do it again, the first aeroplane parachutist replied, 'You can stick your parachute up your <word rhyming with Jannus>.'

Parachutes were to be required wear for many aviators – except for British First Global War pilots – but more of that later.

AN EX-PILOT

We couldn't end this interesting* chapter without including this tale. An American newspaper put up a prize of $50,000 for the first flyer to travel from one American coast to the other, in under thirty days. Step forward, Calbraith 'You can call me Cal' P Rodgers. He was a big strapping lad who didn't strap in, thinking it was better to be thrown clear of a crash than remain attached to the aircraft. He also threw caution to the wind by flying chomping on a cigar.

In New York on a date in 1911 the courageous Cal took to the air in his Wright brothers Model EX biplane, named *Vin Fiz* after his sponsor's new grape drink. The plane had 'The ideal grape drink' painted on the underside of the wings. Anyone seeing Rodgers pass overhead would think, 'Hey, that guy's in a plane. I'll buy some grape drink.' Seems marketing isn't rocket science.

* **Lengthy. - Ed.**

Cal had been taught to fly by the Wright brother who had been involved in the first aircraft fatality in 1908. That's not something you see highlighted on a learner car, is it? 'We will pass you. Or crash you.' Perhaps he spent too much time on this side of things as Cal experienced nineteen crashes on his way out west. He flew into a barbed-wire fence. He clipped trees trying to avoid onlookers. A tyre was burst by a cactus. He was hit on the head by a petrol tank. He hit a rainbow. If it was there, Cal'd crash into it.

Another unforeseen hazard was the crowds who thronged to see him and who thought it a good idea to take souvenirs off his aircraft when he was about to take off. One woman said, 'But this plane's got plenty bolts, how will it miss one?' And they say we're dumbed down nowadays.

Cal reached the official end of the challenge at Pasadena on the forty-ninth-day mark. He wasn't close to the prize but he had a cigar. Ah, but committed Cal wasn't finished. He just had to reach the actual Pacific and, a few days later, he took off. And crashed again – 10 miles short of his target. His *Vin Fiz* was finished. Or was it?

No!

Nothing was going to stop this man, even two broken legs and a fractured collar bone. A month later Cal reached the west coast, eighty-four days after leaving the east one. He did it in style, landing on the beach and taxiing right into the surf at Long Beach. He had done it! Sixty thousand spectators witnessed his moment of agonising triumph. ~~Trigger's Broom~~ *Vin Fiz* only had a few remaining original parts: the rudder, some bits of wings and the lucky rabbit's foot, but what a plane it was.

Cal was not able to enjoy his fame, as a few months later a gull got stuck in his rudder, causing him to plunge into the Pacific. It was only half a mile from where he had triumphantly taxied into the water. He really had no luck.

FIVE

THE FIRST WORLD WAR OR WHEN ARE THOSE PARACHUTES ARRIVING?

In 1914 events too complicated to research took place that led to the global conflict called the First World War. Initially it was termed the 'Great War' and for those selling uniforms and shells it really was terrific. Those who were closer to the sharp edge (literally) called it the Muddy Awful Conflict and they had a point. The one place that was relatively free of mud was the air but not all countries had what you would really call an 'air force'. Especially Britain.

In 1910 the British Army tested aircraft to see if they could be useful. They found that pilots flying over the enemy could see them and note their position. This could have huge military benefits, they thought. However, there were some fears: these noisy engined things might frighten the cavalry's horses. This is actually true.

In 1912, as we'll remember, a competition was held to determine which machine would be the best for the armed forces. Some tried sticking wings onto horses in an attempt to keep things on a familiar horse-based basis, but this resulted in the horses flying off, never to be seen again. So, aeroplanes it was.

In the end the British went for the B.E.2. It wasn't very fast, couldn't climb very well and didn't carry much in the way of armament. But it was stable. This meant that when it was shot down, it crashed without much fuss or nonsense. It was the aeroplane equivalent of the stiff upper lip and perfect for an army that chose its officers on how well they could pass a sherry decanter.

When the Hellishly Dirty War broke out, the British Royal Flying Corps sent its mighty inventory over the Channel to stop the German forces. It consisted of a collection of types that would grace the Shuttleworth museum. They were a mixed bunch, including the Blériot monoplane, similar to the one Blériot flew across the Channel in the other direction five years before. Most were the B.E.2, described as being of a 'tractor' design, so named because they were more suited to agriculture than aerial combat.*

The Germans on the other hand had many, many planes, more than the British and French added together. Because we love German words, this seems the best place to include ones like *Feldflieger Abteilungen*, *Kampfflugzeuge* and *Jagdgeschwader*, which we found in a book and have no idea what they mean. They just sound impressive.

The Royal Flying Corps were there to spot the enemy as part of this new-fangled 'aerial reconnaissance' malarkey. The trouble was when they shouted down 'They're over there' and pointed furiously towards the enemy positions, no one could work out what the hell they were saying or who or what was where. So they invented a reporting mechanism whereby the observers would land, type up a report (in triplicate: white copy to be sent, green copy for filing and pink copy for internal distribution), ride a horse over to HQ, give them

* **No, it's because the propeller is at the front and 'pulls' the aircraft forward. – Ed.**

the white copy, fill in the form indicating safe receipt of the reconnaissance report, file that, then await instructions before returning to their squadron. It wasn't the quickest or the most efficient process.

Heads were scratched as to how information could be gotten more quickly. Some proposed writing down their report (not in triplicate) and dropping it over the fuselage's side so it could be picked up. This was tried but they were either lost or shot at by troops thinking they were bombs.

A bright spark piped up: 'Why not use radio?' They were chastised by higher-ranked officers. This was a war and not the place to listen to music hall tunes, you stupid boy. The bright spark tried again while still smarting from being called stupid. He wasn't the stupid one, the big stupid officer with his stupid face was stupid. Lieutenant Bright Spark was told to stop muttering. He piped up again: 'Why don't they use the radio to send down messages from the aircraft in the air?'

This they did, although it was not without teething problems as the receivers thought the airborne watchers had been taken over by Martians, until informed they were sending down Morse code and not voice messages. This was a revelation and allowed instant communications to be received.

The early sets were so heavy, no observers could be taken up to operate them, so the pilot had to do all the observing, flying and then sending of messages. It's a bit like patting your head, rubbing your stomach and sending a text at the same time. Not easy unless you are an octopus, which few pilots were in the early stages of the war.

When the trenches were dug, it was thought a good idea to get maps showing where they were. It was difficult to sketch them successfully, as the observers kept dropping their pencil sharpeners over the side. Cameras were brought in. And then straps, as these kept getting dropped over the side too. This new 'photo-reconnaissance' was to result in thousands of photographs of trenches. Friends and families would be bored rigid having to sit through a lantern slide show of what snaps had been taken recently. Oh, here's one of the trench at Abbeville. And then here's one of the supply trenches at Mert sur la Ville. And so on until the snoring prevented anyone hearing anything more.

TROUBLE IN THE AIR

It was great having aircraft spotting the enemy but then it turned out the opposing generals didn't like this and so they sent up their own aircraft to ask the British (and French who were also there seeing as it was in France) to stop. The British pretended they couldn't understand and so held up two fingers to indicate trying a second time. The Germans took umbrage at this. Graphic physical gestures were made, along with taunts written on bed sheets that used language not suitable for a family history.

Eventually someone high-ranking said enough is enough and that this was no way to carry out a war. So, the next time they went up the pilots and observers went tooled up. Pistols, rifles and a thesaurus of insults were stashed in the cockpits.

Now, as anyone who's been on a team away day will know, shooting a target on the ground is quite hard. Shooting a moving target that is a small aeroplane going more than 100mph from another small aircraft also going at 100mph while being buffeted by wind, limbs shaking with fear and wearing a thick fur and leather suit is also quite hard.

In order to get more bullets heading towards the target, another bright spark suggested using machine guns. One of the problems was that every time they fired them off, the propeller ended in bits. This posed issues. Another bright spark said they should only use the 'pusher' type of aeroplane. This was different from the tractor in that the engine was at the back. The gunner would have a clear field of fire forward. Unfortunately, when they bolted on the machine gun, it made the aircraft too heavy. To save weight it was proposed that they didn't load ammunition. The gunner could just shout 'rat-tat-tat-tat' at the opposing plane. This was judged to be completely pointless. How would the enemy hear the gunner in among all that wind and engine noise? And what if they didn't speak English?

When the pushers were given decent engines, they could fly with a decent gun and, if the gunner didn't fall out, they stood a reasonable chance of hitting something, as long as the gunner remembered to open their eyes. The first aircraft shot down in the war was a German Aviatik, sent Earthward by a French Voisin pusher in the autumn of 1914. They'd done it!

Now, the pushers were all very well but, like flares and diphtheria, they went out of fashion. The tractor-type machine was now the *soup du jour* but it had that problem: how to get the bullets past the propeller? One idea mooted was to mount the machine guns at an angle out from the fuselage, because it wasn't hard enough shooting at planes already, was it? A new innovation was brought in by Roland 'Remember Me' Garros. He had metal deflector plates fitted onto his propeller so that he could blam away with his forward-firing guns willy-nilly. Unfortunately, one day he had to crash-land and before he could put his plane through a shredder, the Germans had a look. They scratched their chins. Was this a feasible way of doing it?

A Dutch designer working for the Germans had a better idea. Anthony 'Thought the Dutch Were Neutral?' Fokker came up with a clever idea that was to revolutionise aerial warfare. He manufactured a mechanism whereby the machine gun only fired a bullet when the propeller was out of the way. This mechanism meant that machine guns could be fitted pointing straight out the front. All the pilot had to do was line up his machine in the direction of the enemy aircraft, pull the trigger and blammo!

It was to prove deadly. Fokker Eindeckers (One-deckers) ruled the skies with their forward-firing guns. The Allies called it the 'Fokker Scourge' but the Germans thought it more of 'A Good Time'. The Bloody Nightmarish War was to prove one thing: technology would be developed throughout as each side sought to get an advantage over the other. More advanced types of machine came along, such as the Sopwith Pup, the Sopwith Camel, the Sopwith Snipe and the Sopwith Triplane. It wasn't a monopoly – that was a board game about owning property – other manufacturers made aircraft too.

These machines were no good without one thing: petrol. And oil. Oh and water for the radiator. And don't forget bullets for the guns. But most importantly they needed brave people to clamber up into a war machine with the armour plating of a tea towel. Some of these were good at it and became immortal names. They were The Aces.

THE ACES

Aces were pilots who had shot down five or more aircraft, preferably the enemy's. They were a new breed of airperson: the fighter pilot, flying a specific type of aircraft called fighters that were nimble enough to maneuv ... manoover ... manouerva ... zoom around to shoot down other planes. When fighters of opposing sides met, they would have what became known as a dogfight, although no dogs were hurt in the making of this type of warfare. Steely-eyed pilots would swoop and dive and then pepper their opponent with bullets. That was the plan anyway. Oftentimes they lost control or pulled the wings off through over-stressing the airframe or span into the ground all by themselves. The Aces got very good at creeping up behind unsuspecting aircraft and blasting them with bullets. It was thought a bit unchivalrous by those who called women 'maidens' and worried about being roasted by dragons, but aerial war, like that on the ground, is hell.

Britain's Aces included Albert 'Ace' Ball, Edward 'Mr Ace to You' Mannock and James 'Acey' McCudden. France had their Aces too: René *'Le Ace'* Fonck, Charles *'Je Suis un Ace'* Nungesser and Georges *'Un Ace Aussi'* Guynemer. (Sadly, and ironically, Roland Garros, despite his interest in tennis, never became an 'ace'.)

Germany, of course, had its own Aces: Ernst *'Der Ace'* Udet, Werner *'Ich Bin Ein Ace'* Voss and Oswald *'Ich Bin Das Beste Ace'* Boelcke. Boelcke came up with some rules for air-fighting that would be adopted by fighter pilots for years afterwards:

1. Shoot down enemy planes.
2. Don't get shot down by enemy planes.
3. Wear clean pants in case of hospitalisation.

Another Ace we should mention was Max 'Factor' Immelmann. Max was known as the 'Eagle of Lisle' because of his scaly feet and for shooting down many Allied planes. He himself was shot down in 1916 but before this he perfected a combat manoeuvre that was named in his honour: the Immelmann Combat Manoeuvre. It goes like this. The pilot is flying along, then pulls back on the control column, which makes the aeroplane go upwards. It keeps going upwards until the pilot and his machine are upside down and all the change is out of his pockets. He then 'rolls' the aircraft right-side up and flies on, now going the other way. This simple but effective move is still used by aerobatic planes but not as much by domestic airliners.

The daddy Ace of them all was Baron von Manfred '*Ja Ich Bin Der Größte Ace*' von Richthofen. Manny encouraged his squadron pilots to paint their aircraft bright colours. He had his Fokker painted all-over red, thus earning him the nickname 'The Red Painter'. With their coloured machines, his unit quickly also gained a name: 'Monty Python's Flying Circus'. They were given the freedom to move wherever they fancied. Some suggested Provence looked good this time of year, until gently reminded they were there to shoot at planes, not go paddle boarding. So, shoot they did. Richthofen himself downed eighty Allied aircraft in the war, the highest number by any pilot. He didn't get to enjoy his fame, being shot down himself just before he died.

BOMBING IT

In 1911 an Italian officer, Lieutenant 'Giulio' Gavotti, dropped four grenades onto a Turkish camp in Libya. It wasn't a college prank or a TV gotcha but the first-ever aerial bombing. Italy was at war with the Ottoman Empire and one day Gavotti was flying his Taube (German for 'dove') monoplane when he spotted the enemy camp near an oasis. With the cool head of an aerial attacker, he seized his opportunity and dropped his grenades over the side. Whether they hit anyone is not known but, for those below, the dove was no longer a symbol of peace but of aerial war bombing.

Airships, being much bigger, could deliver a lot more bombs onto a target. First used in the Premier World War, they were a right menace and very difficult to counter. While an airship is big, the sky is bigger. If they fly at night, well it's like trying to find an inflated needle in a dark haystack the size of the sky.

Some pilots did brave this night sky to go up against these quaintly named 'baby killers', but their bullets didn't do much. If you can imagine hitting a giant flying elephant full of gas with a peashooter, it would give you some idea of the task at hand. Some clever bright sparks made new types of bullets that exploded on contact and they had the desired effect. Zeppelins were no longer as unshootable as before.

But it was the area of aeroplane bombing that was to see big advances. In the early days of the war, pilots and observers still dropped 'munitions'* over the side of their aircraft. It could be grenades, artillery shells, small bombs or flechettes.** As the weight of the munition was limited by how much a human could lug over the side, it was thought much more sensible to mount them (the bombs, not the crew members) on the underside of the plane. This was OK as long as the crew remembered they were carrying bombs only inches underneath their undercarriage when coming in for a forced landing.

There was an accuracy problem when dropping the bombs as there was no way of aiming, what with the crew having their eyes wedged tightly shut (aerial war is scary as well as hell). Bombsights were developed to help get the weapon in the intended postcode of the target. To help with the accuracy, pilots flew at low level but this brought them within range of the enemy's guns, which could bring them down even lower. Six feet under.

Aircraft that were built to carry bombs were given the title 'bombers', which makes sense. One of the biggest was

* **Military word for 'bangy, hurty things'.**
** **Whatever the hell they are.**

the Handley Page O/400 – the 'O' standing for the sound people emitted when it came within range. It could carry 1,800lb of bombs. Which is a lot. An even bigger bomber was the V/1500 – the 'V' standing for the finger gesture people displayed when it came within range. It was bloody massive, designed to fly to Berlin, drop its bombs and come back. Sadly for those who built it – but unsadly for those who lived in Berlin – the war ended before the behemoth bomber could be used in action.

Speaking of Germany, the Germans were also big into building big bombers. They had Gothas, which were biplanes, bi-engined and by the Holy were effective. They could carry over 4,000lb of bombs to someplace 400 miles away, which brought London within range. They caused so much devastation and death it caused a bit of embarrassment for the British Royal Family, who changed their name from the slightly Germanic Saxe-Coburg-Gotha to Windsor. True story.

Although known for airships, Zeppelin also made winged aircraft. These came in the shape of big machines like the Zeppelin-Staaken R.V, the Zeppelin-Staaken R.VI, the Zeppelin-Staaken R.VII, the Zeppelin-Staaken R.XIV, the Zeppelin-Staaken R.XV, the Zeppelin-Staaken Really Big Bomber Plane* and of course the Zeppelin-Staaken R.XVI.

But these were all beaten in size by a real biggie: the Siemens-Schuckert R-VIII. Its wingspan was 157ft, enough of a distance to hold more than half of the Red Arrows in close formation, not that they'd allow you to do that. They're very precious about their jets, are the Arrows.

The R-VIII had six engines in an engine room like a ship's. Full speed ahead, aye aye Captain and all of that. Luckily for those conscripted as aeroplane painters, only two were built before the end of the war. They sure used up a lot of emulsion.

* **Not really, just checking you were still with us.**

Oh, another one. The Russian Ilya Muromets had brought a new innovation into the front line: four engines, an achievement duly earning its place in the *Vodka Book of Records*. The Muromets came in several different types, such as the 'Beh', 'Veh' and 'Geh'. There was no 'Meh' though because if this mighty machine was a cake on a TV baking show, it'd be a show-stopper. It was the biggest plane on the scene at the time.

The Muromets was dreamed up by a person called Igor Sikorsky. He had originally designed it as a luxurious passenger plane. The Tsar (of Russia, not those modern-day ones that look after drugs or litter picking) had come to see Igor's work and gave him a very expensive watch as a thank-you. It pays to keep in with the rich. But when war came, the airframe was redesigned as a long-range bomber. An innovation was having a tail gunner, which proved to be a thorn in the side and bullets in the front for any attacking fighters.

HEALTH-AND-SAFETY NIGHTMARE

As you'd expect, not much attention was given to Health and Safety in the war. It was full of accurate artillery fire, sneaky snipers, ghastly gas attacks and suchlike, taking place in a trip-hazard-full environment of flooded shell holes. On top of that, it was unsanitary, with shower facilities almost non-existent (unless it rained) and very little in the way of personal protective equipment beyond a steel hat and a thick woolly jumper.

In the air it wasn't much better.

Armour plating was too heavy to carry for most machines and British pilots and observers were not given parachutes. Their commanding officers felt that if they were issued, it might encourage crews to abandon their machines at the first smell of danger. It was seen as far better – more British – to have men plummet to death in a fiery spiral than gently descend by silk parachute to live and fight again. The Germans had them. They weren't as daft.

As we might have mentioned above, at the start of the war some pilots carried revolvers to use on the enemy. Later in the war some pilots were still taking them aloft, not in case their machine guns failed but to use on themselves if their plane caught fire. These were grim times. Nowadays the nearest we can get is flying with a low-budget airline and getting stranded on the tarmac for eight hours without a complimentary bottle of water, but you wouldn't want to shoot yourself. The person who owned the airline however ...

The next time you feel like complaining about your job, about having to sit through interminable team meetings or patience-testing away days, think about balloon observers. These poor persons were lifted up in a wicker basket beneath a big balloon that was anchored to the ground. These static observation balloons gave their occupants the chance to practise bladder control, with them being airborne for twelve hours at a stretch. These S.O.B.s were highly visible and easy meat even for a pilot who found it tricky to achieve the airborne equivalent of hitting a cow in the backside with a banjo. But all was not gloom for the observer, as they were issued with parachutes and, if attacked, could leg it over the side and descend safely to Earth. Try that the next time you're at a strategic planning day.

NAVAL GAZING

Naval officers could see the advantages in aircraft the same way army officers could. With an aeroplane, admirals could get a bird's-eye-view of the battlefield, or 'sea' as they called it. As you'd expect, German naval people were keen users of military airships. They could stay up for much longer than winged aircraft and had their own toilets, but they weren't very cool. Swings and roundabouts. They were also not very good at being nimble, so winged aircraft had to be looked at.

Attempts were made to fly aircraft off and then onto warships. Now, aeroplanes, even the small ones they had in those days, normally require a decent length of flat area from which to take off and land. This presents a problem as destroyers, frigates, cruisers, battleships, dreadnoughts, minesweepers, etc. are not blessed with a lot of this space. Some clever solutions were sought. Structures were built on top of ships' guns to allow a plane a one-off chance to take off. This led to resistance from officers, who complained these new-fangled structures were restricting their guns' movement. Which was a fair point but maybe not that collegiate in attitude.

Another method saw aircraft being towed behind ships. Not sure why. It must have given the sailors something to do. One idea that caught on was to attach 'floats' onto the aircraft to prevent them sinking. These 'float planes' were lifted on and off ships by crane. So that was another thing to keep the sailors occupied.

But all these were just mucking about really. What was needed was a decent landing/taking-off area on a ship. With this in mind, HMS *Furious* had a flat deck built on it. It wasn't ideal as the plucky pilot had to negotiate his way around the ship's superstructure to reach the deck. It's a bit like going to the kitchen for a glass of water during the night and trying to remember where the Lego* bricks are on the floor.

On a notable date in 1917, British naval officer Squadron Commander 'Edwin' Dunning brought in a Sopwith Pup to attempt a landing. If successful, it'd be the first-ever landing

* **Other plastic brick-based construction toys are available.**

on a moving ship. With aplomb he brought the small biplane in and, as if he'd been doing it all his life, brought it for a safe landing. He'd done it! A few days later he did it again but his engine conked out and so did he, his plane going over the side. But the courageous commander had earned his rightful place in the *Sailor's Rum Book of Records*.

Further developments were made for these aircraft-carrying ships that resulted in the boats we know nowadays as aircraft-carrying ships. They had large decks long enough and wide enough for aircraft to take off from and then crash back onto if the pilots weren't paying attention. This is unfair. Landing on a moving ship is not straightforward. The deck – being part of the ship – goes up and down and side to side if the water is not smooth. And if you do make it down, there is always a risk of over-braking and being tipped onto your nose. This can easily be replicated on a pushbike by going fast, then selecting only the front brakes.

To help stop aircraft, 'arrestor wires' were strung across the deck. A hook was bolted onto the aircraft and the skilful pilot flew close enough to the deck that the hook connected with the wire and things came to a sudden, shuddering halt that tested the adhesive in the naval aviator's false teeth. The ghost of Commander Dunning gave a wry look before floating off.

AMERICA FLIES IN

In the last bit of the war the Allies received a boost with the arrival of The Americans. America had taken its time before fully committing to being in the war, and it was criticised for this, but those critics should remember that wars can result in injury and death to those taking part, so any hesitancy is perfectly understandable. You know when a friend asks if you want to go to their place for a barbecue but, even though you love hotdogs, you heard that last week someone else that went to one was sick for three days afterwards due to a dodgy drumstick? That's what entering the First Godawful War was like.

Before they were officially in the war, air-minded Americans had formed the Lafayette Escadrille, even though none of them were called Lafayette nor Escadrille. They were called Norman Prince and Edmund Gros. The latter had also started up an ambulance unit, showing he knew where his bread would be buttered. The American pilots had great names such as Eddie '12-String' Rickenbacker, Kiffin 'Norman' Rockwell and Eugene 'Traffic' Bullard. The squadron was good at shooting at the Germans and earned the respect of its French counterparts, oftentimes hearing the phrase '*c'est bon*'.

The war ended in 1918. The world after the war would be a different one, but it would still have planes in it. Lots of planes.

Simon the Magician's fall from grace was also from the sky. ('The Fall of Simon the Magician', from *Les Tableaux de Rome: Les Eglises Jubilaires* (*The Paintings of Rome: The Churches Jubilee*), plate 4, 1607–11, by Jacques Callot)
(The Met. The Elisha Whittelsey Collection, The Elisha Whittelsey Fund, 1959)

Just a flying woman riding a horse under a balloon.
(Library of Congress, Prints & Photographs Division, LC-DIG-ppmsca-02492)

The Aerial Steam Carriage got as close to flying as a low-budget airliner gets to the city on the ticket.
(Library of Congress, Prints & Photographs Division, LC-DIG-ppmsca-02570)

Who can resist a big fluffy dog? Not one of the Wright brothers – it's his.
(Library of Congress, Prints & Photographs Division, LC-DIG-ppprs-00738)

'I don't know how I struggle to get life insurance.'
(Library of Congress, Prints & Photographs Division, LC-USZ62-48803)

While it was reassuring to have a gunner, it didn't give the pilot the best view in the world.
(Library of Congress, Prints & Photographs Division, LC-DIG-ggbain-15770)

A good landing is one you can walk away from. Or jump down from.
(Australian War Memorial E01882)

think the pre-theatre menu
›oks pretty reasonable.'
Australian War Memorial
01650)

›n airship flying along, not
rashing or burning.
Nationalmuseet, Denmark)

During 1940, the London Planespotters' Group's notebooks were filled with Junkers, Heinkels Dorniers and Messerschmitts.
(National Archives photo no. 306-NT-901B-3)

The famous Spitfire. They won the war, you know.
(Norman Ferguson)

The Hurricane's lack of reverse gears was a pain to the ground crew.
(Norman Ferguson)

A Lancaster. They also won the war, you know.
(Norman Ferguson)

The Douglas X-3 Stiletto. You could take someone's eye out with that.
(NASA Photo E54-1228)

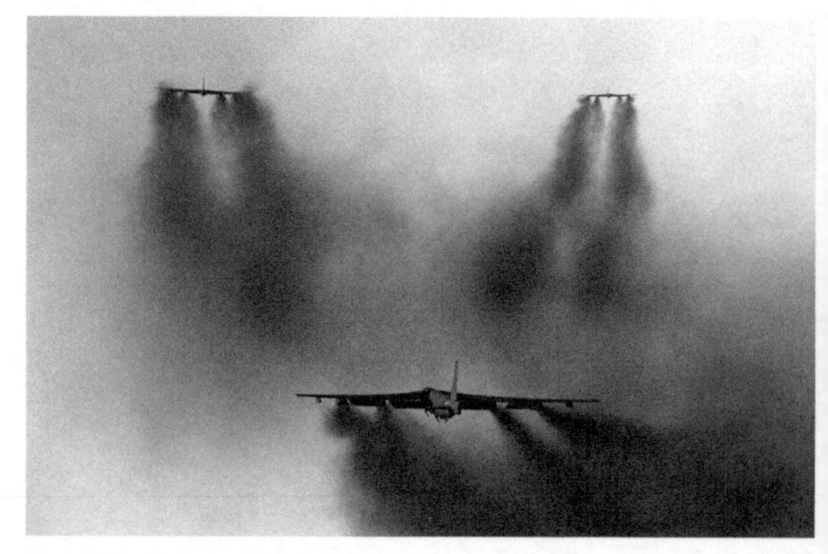

The USAF's 57th Strategic Climate Changers Squadron in action.
(National Archives photo no. DFST8712398)

There were more MiG-21s built than you've had hot dinners (if you're under 31 years old and had on average one hot dinner a day since birth). *(Norman Ferguson)*

Despite its name, there are two Air Force Ones. *(Norman Ferguson)*

Concorde. One of the great British icons. Built jointly with France. *(Norman Ferguson)*

A helicopter. That's it. That's the caption.
(Norman Ferguson)

Even with all the hi-tech gear, F-35 pilots sometimes need a directional steer.
(US Navy)

SIX

THE GOLDEN AGE OR GPS WOULD BE A GREAT THING RIGHT NOW

It is held by those who know about these sorts of things that the period after the First and before the Second Big War is described as the Golden Age of Aviation. Did the bold aviators wear gold jewellery? Did they win gold prizes? Were their machines painted in warm yellow tones associated with that precious metal beloved by people such as Auric Goldfinger? Some of these might be true but it's really used to describe, in general terms, a Very Good Time.*

'But why?' we hear you ask.

Well, at the end of 1918, Britain's air force, as well as other air forces, had a lot of aircraft and aircrew. The British RAF (RAF) had been formed during the war by merging the army and navy air forces. The navy made grumbling noises, not because they were hungry but because they weren't happy. Eventually they were able to break away and have their own

* **Wikipedia editors.**

air force, called the Fleet Air Arm. If they had gone for 'Air Force' then there might have been some FAF-ing around.*

So, this RAF was huge, the biggest air force in the world, with over 20,000 aeroplanes and 300,000 persons. But there really wasn't any need for warplanes any more. Pilots and observers were given their marching orders – literally – and were out of a job. That's peace for you. It's not all good news.

With so many trained aviators around, they had to find something to do. Delivering takeaway food via penny farthings didn't appeal to all. One thing they could do was go exploring. Having faced death above the trenches, they could now face it all over the world. Prizes awaited for those who defied the elements and technology was available to explore unflown routes, set new records and not come down in a hostile ocean. One of the first Golden Age events took place not long after the end of the war. The year after in fact.

THE ATLANTIC AIR WAY

The Atlantic is an ocean, the second largest in the world, covering approximately 41 million square miles of the Earth's surface. It is big. And wet. Flying across it was seen as the next big challenge for bold aviators. As was the way in aviation endeavours, there was a prize at stake. The prize at stake was £10,000 from a British newspaper. This cash would go to the crew who flew from northern America to Britain or Ireland, non-stop, in a period of seventy-two hours without disappearing without trace, dying in a fireball or being swallowed by sea serpents.

On a famous date in 1919 residents of Newfoundland watched agog as a blue US Navy airship floated into view.

* **Get on with it. – Ed.**

This was the C-5. Keen-eyed fans of 1980s tech-wizard Clive 'Entrepreneur' Sinclair will recognise this name as denoting great technical success. The navy aimed to fly their airship across the Atlantic and it may well have done so, if the wind didn't tear it from its moorings and send it eastwards where it went to Airship Valhalla.

Also in 1919, the US Navy sent three flying boats called Curtiss NCs (but everyone knew them as Nancies) on their way across the cold, wet ocean. They couldn't win the money as they wouldn't be able to do it in the seventy-two hours stipulated by the prize awardees. (There were allegations that the conditions of the British-originating prize were altered to make it impossible for the Americans to win. It was also alleged that any winner had to be able to pronounce 'Scone' properly and know what knife to use with fish.)

The plan was to fly via the Azores and then Portugal, then Britain, before reaching mainland Europe. The navy flyers hoped the majority of the route would be over the Atlantic and not in it, but the epic journey was not plain sailing. The big flying boats flew into fog and became separated. One landed on the sea. Another landed on the water but taxied to the Azores. One did make it all the way to Britain, arriving in Plymouth, where a welcome committee said, 'I hope you've booked, we're very busy this time of year and we're having fish, is this the knife I should use?' The bold Americans had done it, albeit in twenty-three days, approximately twenty days too late to be in contention. 'Better late than never,' said some British people, remembering the recently finished war.

Still in 1919, two bold British flyers wanted to win the day, the dosh and avoid a cold and wet ocean fate. Harry 'Harry' Hawker and Kenneth 'Mackenzie' Grieve clambered into their Sopwith Atlantic. Was the name tempting fate?* They would soon find out. The Atlantic was a biplane, which was good as the two-wings-better-than-one diktat still held firm. It had one engine and, while that was better than a glider, it held inherent risks as a single point of failure. Hawker was a test pilot for the plane-making company Sopwith and not Hawker as you might have thought. That came later.

Harry and Ken set off but faced difficult flying conditions. Then the one and only engine started overheating. Uh-oh. It needed water, but where to find some? They were flying over one of the biggest expanses of H_2O on the planet, but it was as out of reach as orbital mechanics to a Flat Earther. Eventually it ran out of puff, and the Atlantic descended towards its namesake. As they lost height, the crew had time to ponder. Should they have used a seaplane? Or a plane with more than one engine? Why do the good ideas always come too late?

The Atlantic landed on the water near to a ship. The plucky duo were plucked out the water by a ship, which fortunately was heading to Britain. When they arrived, they were feted. Those Brits sure love a failure! The pair were awarded £5,000 as a consolation prize. This was better than what they could have won: an ironic speedboat.

Raynham & Morgan might sound like a paper supply company but they were actually two aviators who wanted that ten grand. Shortly after the Atlantic had left from Newfoundland, R & M made their bid for glory, also from Newfoundland. In their Martinsyde Raymor** they took off to

*** How can you tempt fate? Fate's fate, it's going to happen anyway. – Ed.**

**** A clever name taking the first bit of Raynham's surname and the first bit of Morgan's, although if they'd taken the last bits to form 'Hamorgan' it might have been funnier. Might.**

try to overtake Hawker. We say 'took off' but really they 'bit off' more than their aircraft could 'chew' as it only rose into the air for a very short time before smashing back to Earth, its propeller continuing the mastication theme by biting into the Newfoundland dirt. The shaken Raynham gave a quote to a newspaper reporter: 'Well, that's the end of a perfect day.' He was being sarcastic.

Another British attempt was that of Mark 'My Words' Kerr, who aimed to take a giant Handley Page V/1500 bomber over the pond. Kerr was an admiral, which keen-eyed fans of military rank will know is a navy rank, generally known for travelling on the sea. As it turned out, Kerr didn't travel on anything as the flight was abandoned. The behemoth biplane took too long to assemble and test – and have the right kind of seat fabric chosen after being shipped over from Britain – that it missed its chance of glory. The plane had also tempted fate by being christened *Atlantic*.

At least these ones got to their starting points. One attempt that didn't was by a Major JCB 'Digger' Woods, who named his Short Shirl *Shamrock* (a tricky tongue twister 'tis true). The luck of the Irish wasn't with him, however, as the plane crashed into the Irish Sea while flying to Ireland to begin an attempted east-to-west route.

THE EPIC FLIGHT

This year (1919) was to be an epic one for yet another two flyers in Newfoundland waiting for their chance. The two were John 'Jack' Alcock and Arthur 'Whitten' Brown. Both were British – both had flown in the war and both had been prisoners of that war. They were not afraid of a simple flight across some water. Pah, piff and tush! Or if they were, they never let on.

The two airpersons had travelled to Canada to prepare for what could be a history-etching journey or a descent to disaster. While their machine was carefully checked and fuelled, they pondered the big problem they faced: should it be 'Alcock and Brown' or 'Brown and Alcock'? Before it became a beastly fist-fight, they sensibly went with the alphabetical order, though Arthur thought first names shouldn't be disregarded. There really wasn't much to do in Newfoundland.

With it finally decided, Brown and Alcock pulled on their lucky pants and clambered into their Vickers Vimy biplane. It came with two engines. When Sopwith heard they were about to use such a type of craft, he gnashed his teeth. Why hadn't he thought of that?

The big aircraft trundled out to the take-off area. The Vimy was carrying a heavy load of fuel and, more importantly, the Kendal Mint Cake essential to any venture. They would need all the minty fortitude available as the Vimy had an open cockpit. To add to the fun, the weather was rough. The wind blew off the propeller from the radio's generator so they couldn't get Radio 4. No *Archers*? 'Yay,' said the Glaswegian Brown. It was also foggy. Plus it was cold and their speedometer froze. All this meant they didn't notice they were plummeting to the ocean until it was very, very late and they were very, very close to the sea. When he saw the water getting bigger, Alcock pulled on all his reserves and, more importantly, the control column to bring the nose up, up and away from pointing towards the cruel sea. He strained and strained and, with only feet to go, the big biplane reared up. They flew on, glad the big machine had enough space to store spare underwear.

The hours, like their engines, droned on. An exhaust pipe had burst, thus creating a fearful racket and ensuring communications between the two flyers was made VERY

DIFFICULT. The Vimy had no autopilot and no inflight toilet facilities, so the two just had to sit and not bare it as they plodded on. The Vimy could only do 100mph and, with Brown having forgotten to bring a magazine, the time passed slowly.

At one point Brown clambered out of the cockpit. Alcock was alarmed at this: was his navigator making a run for it with all the Mint Cake? He wasn't. The engines were being covered by ice, and brave Brown had climbed out to get some to chill the champagne because they were nearing the end of their journey. After hours of bum-numbingness, they could see the green, green grass of Galway up ahead. They had done it! Well, not quite. As Shakespeare said, 'There's many a slip betwixt cup and ground.' Brown and Alcock still had to get themselves down without breaking their necks.

Pilot Alcock faced the challenge of doing this. He was tired, exhausted and really needed the loo. He looked for somewhere decent to land. Brown quickly flicked through the *Book of International Airports* under 'I' for Ireland but didn't find any. That field over there would have to do. Alcock lined up for landing, slowly reaching for the ground, gently floating over the golden Galway sands, guiding the big plane slowly down, down, down ... the gallant duo were so close to success they could touch it. And touch it they soon did when Brown's field turned out to be a bog and the giant plane quickly nosed over. It mattered not a jot as they had really done it! Locals came out to see who was making all this mess of their bog, while others more business-focused began printing souvenir commemorative plates and tea towels.

Brown and Alcock were lauded and applauded for their heroically epic journey. They had shown the way across the Atlantic. As they relaxed, toasting each other's success, one thought struck them: where was their luggage?

Their contribution to aviation was rewarded with knighthoods, a statue at Heathrow and a stamp on the fiftieth anniversary. Sadly, Alcock was not able to enjoy his status for long, dying in, guess what? Yup, a plane crash later that year.

THERE AND BACK

Most of the time the wind across the Atlantic blows from the American side towards the European side. It is known as the 'prevailing wind', as it comes before the 'vailing wind'.* This makes it sensible to fly the same way: west to east. To be blown across is more helpful and energy efficient than flying into the wind and therefore a Good Thing.

Just weeks after Brown and Alcock's success, another epic flight was about to get going. It was via an airship with the name the R.34 that took off from East Fortune in East Lothian in east Scotland. Yes, they were going to go against that prevailing ~~view~~ ... wind. On board, alongside the normal Royal Navy crew, were two extra passengers: a crew member who was meant to be staying at home and a cat called Whoopsie. Having stowaways meant extra weight and for a while it looked as though they might be 'dropped off' but cats are so cute, yes they are, cutey puss, aw look at its wee face, so it was kept on board. Although not quite as cute, the human stowaway was also retained.

Like all flights across an ocean, it wasn't a walk in the park. At one point the crew had to use all their bubble gum – not to enter the *Irn Bru Book of Records* for biggest airborne bubble, but to fix a leak. Luckily it held and after four days of flying, the big airship reached Long Island, which is in America. They'd done it! An unusual aspect of the flight was one of the crew jumping out. Not through fear or snapping after being cooped up for days hearing the same sea shanties, but through

* **No it doesn't. – Ed.**

practicalities. On arriving at the landing ground, Major 'John' Pritchard used a parachute to get to the ground and then issued instructions on how to berth the 643ft-long airship safely. It had arrived with just forty minutes of fuel remaining. It should have had thirty hours' worth but, you know, the Statue of Liberty, Central Park, etc., won't photograph themselves.

The crew were celebrated, given a ticker-tape welcome in New York and taken to meet the US President. They then properly etched their names in that book of records by getting back on board the airship and flying back over the Atlantic, thus completing the first return journey. The R.34 flew to East Anglia rather than East Lothian, disappointing those in the East Fortune Planespotters' Group who'd missed the take-off.

This epic achievement is marked with a commemorative stone at its take-off site, which is now an aviation museum with a café serving scones (fruit and plain), tray bakes and paninis. The gift shop is well stocked with a range of aviation goods, including the five-star-reviewed *Little Book of Aviation* by those fine people at The History Press.

MORE EXPLORATION

Much of the early flying was done in Europe and America but flyers had gotten bored of these and wanted to fly someplace new. People all over the world saw huge changes as aviation expanded its reach. Where once the locals had looked out at animals strolling over plains, steppes, deserts and jungles, etc., they now looked out at bits of aircraft wreckage on plains, steppes, deserts and jungles, etc. To those on the ground, seeing powered aircraft for the first time provoked different responses. Some looked up in wonder. For others, it stirred something primordial: did anyone get the registration?

There can be no doubting the courage of flyers who went to places never flown above before, except by birds, bees, bats and other things that don't start with 'b' – clouds, for example. Or wasps. And before this, pterodactyls. If an engine conked out, you faced a crash-landing in a remote location far away from a hydraulic fluid supplier.

These pilots were trailblazers, their aim to venture into the known unknown, or the unknown unknown, or in the case of the [name of low-cost Irish airline name removed for legal reasons] pilots, the known to be far away from the actual city on the ticket. The desire to sit in a draughty cockpit hundreds of miles away from a decent toilet was not unique to Brown and Alcock. Many others queued to pull on their thick leather trousers and goggles and venture into the deep blue yonder.

The same year as two Britons flew across the Atlantic, a couple of Australians attempted to fly across the world from Britain to Australia. Ross 'Macpherson' Smith and Keith 'Macpherson' Smith (yes, they were brothers) covered the enormous distance in twenty-eight days. Although worried they'd get nowhere fast, with death at one's elbow, The Smiths never got into a panic and kept each hand in glove while about to oscillate wildly in turbulence. As they closed in on the prize of £10,000, Smith told his brother to keep calm, saying, 'You just haven't earned it yet, baby.' These charming men had done it, though. When they arrived, the Sheilas asked them to take a bow. What difference did their epic flight make? Nothing immediately – these things take time – but it was a rush and a push in making the world just that bit more connected.*

Typical of the period, both of these epic voyages were followed by tragedy. Ross lived a couple of years longer than

* **That's plenty of Smiths references. That joke isn't funny anymore. – Ed.**

Alcock before he died in a plane crash. Not that we wish to add to the World's Weirdest Plane Facts on those channels that have a lot about Nazis and UFOs, but both died flying a Vickers Viking. Spooky? A jinxed plane type? Or just a pretty mundane coincidence? We just don't know.

Around this time, there were many other epic flights, too numerous to ~~research~~ detail. They included:

- Paris to Tokyo
- Tokyo to Paris
- Paris to Timbuktu
- Paris to Tehran
- Paris to Basra
- Paris to Bandar Abbas
- Paris to Jask*
- Rome to Rome (via Australia, Japan and China).

Another epic flight was from Britain to South Africa. (This was a popular route as it doesn't involve those jetlag-inducing time-zone changes.) The pilot was Sir 'Alan' Cobham, whose flair for flying and publicity earned him great attention for this and other flights. He had flown to Rangoon and back (it was only when he arrived in the Burmese city he remembered he'd left the gas on). He also flew to Australia. When he returned from the land down under, he landed on the Thames in front of the Houses of Parliament in a blaze of publicity and not aviation fuel as the landing was safely carried out.

For something to do, Cobham flew around Africa and then flew around Britain, organising air displays. He wanted to promote the advantages of aviation to the British public and gave many their first flight, as part of 'Monty Cobham's Flying Circus'. Cobham thought that, one day, travel between

* **No idea.**

Scotland and London would only take a few hours and that there would be as many airfields as golf courses. You only have to fly over East Lothian to realise he must have been sniffing the avgas as there are so many courses there you can play from the North Sea to the River Forth without going out of bounds. Cobham went on to develop techniques for refuelling aircraft in the air, which became standard operating procedure for pilots who wanted to have more fuel.

The Smiths continued to make records. The biggest ocean in the world is the Pacific and it would take an epic flight to fly it. Charles 'Kingsford' Smith, Charles 'Smith' Ulm and others not called Smith figured they would be the ones to do it. In their Fokker Trimotor, *Southern Cross*, they set off from America. It couldn't be done in one go – the Trimotor didn't have the range. The clean-shaven Smith led the way (he got to sit up front) and nine days later landed in Australia. They had routed via Hawaii – what people will do to get a colourful shirt. Smithy went on to be the first to fly across Australia non-stop – bush landing fees were extortionate – and the first to fly from Australia to New Zealand. Like many pioneers, he died peacefully in his bed. Not. He went missing while attempting another record flight, from Britain to Australia in 1935.

One man who wasn't 'posted' missing was Wiley 'Coyote' Post. Post, along with navigator Harold Gatty, flew around the world in eight days, fifteen hours and fifty-one minutes, in 1931. This was a new record. He loved doing it so much he decided to do it on his own, two years later. He used an auto-pilot he'd developed but Gatty didn't take it personally. Post inevitably died, after crashing in a lake, a few years later.

Throughout these flights Post had one eye, not on success or the fuel gauge, but on everything, having lost the other in an industrial accident that gave him the money to buy his first airplane. That is making a silk purse from a sow's ear right there.

Now, while the Atlantic had been crossed in 1919, some said pedantically, 'Yeah, but it was only the North Atlantic.' The South Atlantic was still unflown over. Would this remain so after 1922? Commander Artur 'de Sacadura' Cabral (like his aircraft, neatly trimmed) and Admiral 'Carlos' Coutinho (clean shaven) hoped not. On a date in that year the pair took off from Lisbon, Portugal, in a Fairey IIID, a single-engined biplane seaplane. The aircraft was given the name *Lusitania*, which we suppose is better than *Titanic* or *Exxon Valdez* but only just. It did follow its naval predecessor by sinking after a float was damaged on landing. Another aircraft was shipped out smartish. And this was lost after engine failure. Third time lucky? Why not. The intrepid duo reached South America seventy-nine days after departing Portugal. They had done it! It was a great feat of navigation, although, to be fair, South America is quite big.

As we know, the Americans had taken their flying boats across the Atlantic. In 1924 they wanted to go further. They wanted to fly around the globe. Could it be done? Maybe. Four Douglas World Cruisers were prepared. Presumptuous names? Time would tell. The four planes were individually named, but sounded like a 1980s soft-rock reunion tour. They were *Seattle, Boston, Chicago* and *REO Speedwagon*. Things didn't get off to a great start as one crashed into a mountain a few weeks into the trip. The remaining three continued across Asia and Europe until another was lost in the Atlantic. Eventually, after five months, the remaining two plus a substitute made it back to Seattle. Wait, a substitute? Yes, the pilot who had ditched in the Atlantic got in another aircraft in Nova Scotia and joined the rest. Is this not a bit like hiding in a shrubbery during a school cross-country run and then rejoining for the final lap, Simon Jones? We're not sure but it was a terrific achievement all the same.

LINDBERGH STAYED UP ALL NIGHT TO GET CALLED LUCKY

There's an expression that goes, 'There are old pilots, there are bold pilots, but there are not many pilots named "Daredevil" who die in their beds.' One person was to prove this aphorism.* His name was Charles 'Daredevil' Lindbergh; he was given this insurance-premium-raising nickname due to his keenness as a barnstormer. Chas was mad keen on doing wing walks, which are exactly as they sound – if you stay *on* the wing. Wing fallers tended to only do this once.

Luckily, Lindbergh gained a new nickname: 'Lucky', which came as a great relief to his insurance broker, friends and family. He still pushed his luck, did Lucky, parachuting four times from mail planes he was flying. One day Lindy was flying on a mail run from St Louis to Chicago when a thought struck him: did he leave the gas on? No, he'd definitely switched it off. Then another hit him: if he was going to have to sit in a bone-shaking aircraft for hours on end, why not arrange it so that he would be arriving in Paris, France, rather than Chicago, America? Then another thought hit him: had he locked the back door? These thoughts occupied him for a few more bone-juddering miles but the one about Paris would come back ...

What brought it back was that old favourite incentive: a prize. The Orteig Prize offered $25,000 to the first aviator to fly non-stop from New York to Paris. In the 1920s this was a lot of money. It is also a lot of money nowadays if you find it down the back of a sofa. It was offered by Raymond 'Moneybags' Orteig, a New York hotel owner who was keen to drum up business, but who maybe didn't see the flaw in the plan that encouraged flights *from* New York?

* **Or is it an idiom? Adage? Axiom?**

A few had a go. Former French ace René 'Ace' Fonck was one of them. In his specially designed Sikorsky S-35, the moustached French pilot roared down the runway in Long Island, America – not into the pages of aviation history but into a ball of flame, as the machine failed to leave the ground and overturned. It was a complete disaster, with two of the crew dying in the plane, which might have just been a tiny little bit overweight. It was luxuriously fitted out with mahogany and leather panelling, a couch, chairs and fine wines. To accident investigators this would start alarm bells ringing. Fonck had seventy-five kills during the First Awful War and now he had two after it.

They weren't to be the only ones not to make it. Two American pilots were testing their *American Legion* airplane when, in avoiding some trees, their craft stalled and failed to avoid the ground.

Next up (or down) was another French war ace. Charles 'Ace Aussi' Nungesser chose to fly the Atlantic from east to west as the prize could be won if you flew that way, something that would have spoilt our joke up above.* Chas took off from Paris aboard a Levasseur PL.8 aircraft named *L'Oiseau Blanc* (L'Oiseau Blanc), which had a picture of a coffin (!) on the fuselage. With him was François 'E' Coli, another war veteran. They were seen flying over Ireland and then not seen flying anywhere else. Some people in Newfoundland reported hearing an aircraft but, like those of children seeing Santa at Christmas, the claims were never backed up with hard evidence. The two intrepid flyers were gone. Among other tributes, Nungesser and Coli's contribution to aviation's exploratory spirit for financial gain was commemorated by the obligatory tea towel and lakes named after them in Canada: Nungesser Lake and

* **If you can call it that. – Ed.**

Coli Lake. Not Lake Nungesser or Lake Coli, like in most other countries.

So with these attempts out of the way, the taxiway was cleared for Lindbergh. He had invested some of his own money and that of St Louis businesspersons. His chosen aircraft was built by Ryan. Not Ryan that Simon knows from work – you know him, tall chap, good with engines; his sister worked at the garage for a while. It was a company called Ryan.

When Lindbergh first saw their machine, he coughed, pointed at the front of the aircraft and whispered to the designer, 'Where's the front windscreen?' It wasn't there. The pilot – Lindbergh – would have to fly 3,600 miles without being able to see what was in front of him. (Though, if he knew what was coming, he might have welcomed this.) The designer shrugged. Was he French, thought the about-to-be-bold Charles? No. The logic was this: as there were no hills in the Atlantic, he wouldn't need to keep a lookout. What could he run into? That actually did make some sort of sense and, with no option available to put in a windscreen, preparations continued.

The aircraft was named *The Spirit of St Louis*. Fantastic product placement, or at least would have been if they'd thought to market an alcohol of that name. Ah, was Prohibition on then? That would explain it. As you were.

On a date that definitely echoed down the aviational decades in 1927, Lindy clambered into his silver machine that he hoped would propel him into fame and fortune. The fresh-faced flyer's adrenalin was high. He was excited. Who wouldn't be? He was going to Paris, baby! The Eiffel, the Louvre, rude waiters – all awaited him. (If he made it.) The thought of those Parisian delights kept him awake. What was a croc monsewer? Hell, he'd have two! Part of his mind was full of anxieties: what gifts should he get everyone?

Where would he buy stamps? Had he packed his passport? Common worries for anyone going abroad.

Once up in the air, Lindbergh pointed his airplane towards Europe. Then he basically kept doing it. On and on he flew. For thirty-three hours Boldy Prince Charlie battled storms, fog and the realisation he couldn't speak a word of French. Was *chausseux* a horse or a shoe?

He also battled tiredness, having had very little sleep the night before. No, not in that way. He was up making preparations for the epic flight. Mid-flight the plucky person from Michigan was so tired he flew with the window open to keep him awake. Knowing that, if successful, he would be required to make many media appearances, and hearing that salt water was good for the skin, Chas flew near the water so the sea spray blew into his face. A quite extreme skincare regime it has to be said.

Finally, after many, many hours, the audacious American spotted his destination. It was estimated that 150,000 people came to greet him at the landing field at Le Bourget (The Bourget). The airfield owners were upset: why didn't they think to charge entry?

Lindy brought his craft down and managed to not chop any of the adoring crowd into bits with his propeller. He'd done it! Overnight the one-time flying mailperson became an overnight sensation, feted everywhere he went. And he wasn't a fan of a fete. He couldn't stand chutney. But he grinned and bore it. Just as he had to when watching the commemorative plates and tea towels flying out the shops, and him not on a royalty.

His success was hailed the world over. Newspapers headlined with:

Lindbergh Lands in Paris
Delaware Sunday Tribune

Lindy Lands after Epic and Amazing Flight!
Melbourne Post

Lindbergh Comes to London. Will It Affect House Prices?
Daily Express

Lindbergh returned to America by boat. He may have been daft, but he wasn't stupid. Later in life he was associated with fascism; it's said people drift to the right as they get older, though this is maybe taking it a bit far.

WOMEN IN THE AIR

Around this time it was clear it was not just men who wanted a piece of this aviation action. While some males looked down their noses and over their moustaches in a sneering way, others welcomed women into the brothersisterhood of aviators. The women didn't give a flying Fonck. They just got on with it.

Two of the most famous women aviators (or waviators as no one called them) were two whose first names began with A: Amy 'Flyer' Johnson and Amelia 'Flyer' Earhart. They both set records and inspired others with their epic flights. But what did they actually do?

AMY JOHNSON

Amy Johnson is to the UK what Amelia Earhart is to the USA – a woman aviator who pushed the boundaries of what was expected of women aviators. She faced difficulties in achieving her goals. She was from Hull for one thing, which, as its name suggests, is more of a maritime town.

At the time women weren't really thought of as being air-minded but this determined Yorkshireperson eventually won through and achieved both her pilot and engineer licences, which was handy as these early machines weren't the most reliable.

The bold Amy set her sights on Australia as it was hot and sunny and really far away from Yorkshire. She acquired a small biplane, a de Havilland Gipsy Moth, which she named *Jason*, as she loved Greek myths but didn't have enough paint for '*and the Argonauts*'. On a momentous date in 1930 Amy and *Jason* set off on their epic trip. They encountered many obstacles. *Jason* suffered exhaustion while Amy had technical issues, or maybe it was the other way around. At times it was also hot. Desert hot. And that's hot. She also encountered bureaucracy, cloud-covered mountain ranges, sandstorms, broken undercarriage struts, consulate dinners, unseen ditches and heaps of ants. At one point she had to put down on a football pitch and crashed into the goal posts. Back of the net. Not.

But undaunted by these setbacks, she flew on. On arriving in the land down under, Johnson made a spectacular arrival, by crashing during landing and flipping the plane over onto its back. She'd heard the hoary old joke about Australians living upside down and so wanted to endear herself to those watching.

With her name etched in the *Tinnies Book of Records*, the brave flyer continued flying. With her husband, Jim 'Mr Johnson' Mollison, they became the first husband and wife crew to crash after crossing the Atlantic. They also took part in a famous air race to Australia, flying a de Havilland Comet. This beautiful-looking aeroplane was painted black and suitably named *Black Magic*, being delivered by a mysterious man in a polo neck. Sadly it wasn't that magical and the engine went kaput in India.

After the Second Big War broke out Amy joined the Auxiliary Transport Auxiliaries. The RAF wouldn't let women pilots fly in combat but would let them risk their necks delivering their aircraft. In 1941 First Officer Johnson was flying an Airspeed Oxford from Scotland to Oxford that went missing over the Thames, which isn't that close to Oxford. Many theories have arisen about what happened but one thing that is certain is she wasn't around to fly any more. Her pioneering achievements are marked with statues, which can be seen in Hull or Herne Bay as that's where they are.

AMELIA EARHART

Amelia Earhart symbolises the type of person keen to prove that women were just as capable as men of taking off never to be seen again. She first caught the headlines when entering the *Budweiser Book of Records* in 1928 as the first woman to fly over the Atlantic. She followed several other women who had failed to make it: Princess 'Disappeared' Anne of Löwenstein-Wertheim Freudenberg, Frances 'Also Disappeared' Grayson and Elsie 'Yup, Disappeared' Mackay.

Earhart was chosen by Amy 'Be My' Guest, a rich American heiress who was sponsoring an attempt to fly over the big ocean. Amelia fulfilled the criteria of being a pilot, being American, having good manners and being physically attractive. Sorry, what? Yes, things really were different back then. She wasn't going on her own, oh no. A woman, in a plane, all by herself? Heavens to Betsy, she might break a nail or distract the menfolk or something. Oh no, that couldn't happen. She was to fly with two men, who let her sit in the same plane as them. So that was big of them. Off they went, before landing in Wales after twenty hours or so.

This wasn't enough for the intrepid Amelia, who resolved to fly on her own. And in 1932, this she duly did. She took off in a single-engined Lockheed Vega and landed in the same one many hours later. She touched down in Ireland and when out of the cockpit a local asked her if she'd come far. 'From America,' the aviator replied. 'Oh yes, I heard it was an awful place,' replied the local.

Amelia was now a global celebrity. She helped set up a group called The Ninety-Nines for ice cream and chocolate flake enthusiasts. No, it wasn't! It was to promote and support women in aviation. And there were ninety-nine of them. Amelia E was not one for sitting about on her rear undercarriage. She was keen to meet new challenges and none were bigger than the world, and by golly she would fly around it. Her first attempt saw a ground loop* in Hawaii that forced the abandonment of the attempt. She tried again, leaving Miami on a famous date in 1937. With Fred 'Navigator' Noonan, she set off in a Lockheed Electra 10E (the 'E' standing for 'Earhart', as many did when she came into a room).

They succeeded in getting over the Atlantic, then past Africa and Asia, covering over 22,000 miles. The next leg of their journey was over the Pacific Ocean. The two left Papua New Guinea heading for an island called Howtoland Island, over 2,000 miles away. What happened next has become one of aviation's most enduring mysteries. Earhart was heard by a US Navy ship near the island but after a while nothing more was heard. Theories continue to abound about what happened. Did they ditch and sink? Were they shot down by the Japanese? Were they abducted by aliens?

Earhart and Noonan might have ended up on a desert island. Of course, this was in the days before she would be

* **Aviation term for 'crash'.**

asked to pick what records she would have chosen to take with her. It would be very easy to make bad-taste jokes* about which songs she might have picked, like Lisa Stansfield's 'Been Around the World (And I Can't Find My Baby)' or U2's 'I Still Haven't Found What I'm Looking For', but luckily they hadn't been recorded in 1937.

Amelia Earhart is remembered by many tributes. A tree in Hawaii, a student hall of residence, a bridge in Kansas and a lighthouse on the island she didn't reach are just some.

GO WEST

Another epic flight – or rather flights – was that by twenty-four Italian flying boats in 1933. Under the guidance of an Italian person called General 'Italo' Balbo (chin beard), these Savoia-Marchettis flew from Rome to Chicago, going that much-feared east-to-west route. This was a huge undertaking. Passports needed sorting, entry visas had to be applied for, accommodation had to be organised, outfits had to be chosen for the inevitable civic function, etc., etc., never mind the actual flying and navigating.

Their arrival showed the aerial prowess of fascism in a good light. The Americans greeted the Italians warmly, giving them a ticker-tape parade in New York, although some New Yorkers wondered why they were giving each other Roman salutes. Balbo had been appointed by Mussolini as air minister despite him knowing nothing about the subject, something that has never happened in any other government since. The fascist flyer's feat resulted in a large formation of aircraft being known to this day as a 'balbo', which beats a statue, we reckon.

* **Too easy it appears. – Ed.**

GO NORTH

There was an opportunity to etch your name into the *Frozen Daiquiri Book of Records* by being first to visit a pole. The North Pole was more popular for this as it was a lot nearer to most flyers.

The airship *America* had three goes at immortality. In 1906 the unrigid airship was shipped in bits to Spitsbergen, which is really far up in northern Norway. During the pre-flight construction the team, led by journalist Walter 'Yes I'm' Wellman, battled having the instructions blowing about in the wind. It wasn't wholly successful: the engines fell to bits. Walt was frustrated at this, something buyers of self-assembly home furniture will be familiar with. The attempt was abandoned.

In 1907 attempt number two took place. The airship left Spitsbergen and flew into fog and had to be put down on a glacier. The crew were rescued and the attempt was abandoned.

Two years later another attempt was made. Technical problems forced *America* down and the attempt was abandoned. When Wellman heard that another person had gotten to the pole, he went outside for a bit and the sound of wooden objects being splintered could be heard.

But Walt wasn't done with epic flights. In 1910 the cold was replaced with the wet. The *America* had been developed and left America to cross the Atlantic. It might well have done so had the brave balloon not been blown off course following engine problems. The crew got into their lifeboats and the Little Airship That Could flew off on its own, never to be seen again. Although it might seem like a catalogue of failures, the *America*'s last-ever flight did contain one etching into a book of records. The crew had been accompanied by a cat called Kiddo. It caused havoc, chasing a mouse around the cabin, with furniture getting smashed up, them being electrocuted and bashing each other over the head with a variety of objects but always returning to full health the next moment. When one of the crew reached the end of his tether with the moggy aviator he made the first air-to-ground radio transmission to say, 'Roy come and get this goddamn cat.' Whoever Roy was, he wasn't able to take the cat and so it continued on the flight.

Oh yes, polar flights. Knew there was a reason we were here. Roald 'It's not Ronald' Amundsen had etched his place in the *Akvavit Book of Records* by leading a team to be the first to reach the South Pole in 1911. He beat Captain Robert 'Falcon' Scott, who came second but didn't come into good luck, running into bad weather, low supplies and perishing on his way back.

Amundsen wasn't satisfied with his impressive achievement and wanted to try new cold-place epic adventures. In 1923 he attempted to fly over the North Pole, taking a Junkers J-13 monoplane and a Curtiss Oriole biplane fitted with skis, which he named *Kristine*.* Would he be first? No. Roald's aircraft got damaged on the journey north and the attempt was shelved.

A couple of years later the bold Norwegian was back, this time with two Dornier Wal flying boats, with the lyrical names N24 and N25. Both headed north but both had to stop. N24 suffered engine problems – one of them stopped working – and so the plane was abandoned about 150 miles from the hallowed pole. This meant that both crews had to return in the one aircraft, N25. To take off they had to clear a whole lot of snow to form a runway. Then it got foggy. Food was running low. No one could face playing Rock, Paper, Scissors any longer. They all got into the one working Dornier. The

* **The aircraft, not the skis.**

engines were run up to full power. The big machine picked up speed, sliding over the icy wastes. Would they hit an ice bank and be smashed to pieces? Would they be distracted by seeing a polar bear? Was their plane overloaded and so they'd have to have another go? It was the last one. On the second go, they made it off the frozen wasteland, crawling into the air. They turned south and headed for dry land, which they reached. What is Norwegian for 'phew'?*

This wasn't a big enough hint for Roald and in 1926 he was back again. This time he brought an airship called the *Norge*, which is Norwegian for the 'Norway'. While Amundsen aimed to be the first man to have reached the South Pole by dog and the North Pole by air, an American by the nominatively deterministic name of Lieutenant Commander Richard 'I'm Like A' Byrd went and stole a march – and Amundsen's thunder – by flying to the northernmost part of the world first. Or at least saying he did. Byrd and pilot Floyd 'Gordon' Bennett claimed to have reached the pole in the *Josephine Ford* (which despite its name was built by Fokker) but doubts remain that they actually made it.

But no matter the veracity or not, everyone thought the Americans had done it. Those Britons who remembered Scott's feelings on getting to the South Pole and seeing Amundsen's Norwegian flag felt a bit of what the Germans and others called *schadenfreude*. But Amundsen had bought all the maps and didn't want them to go to waste, and so he insisted on going anyway. After lifting off from Spitsbergen, the balloon floated gently over the ice on its way to the fabled pole. Sixty-eight and a half hours later the balloon landed in Alaska. They say if you're not first, you're last. But you could be second too. And then probably first when the others are discounted. They might have done it!

* **Puh. - Ed.**

The pilot and designer of the *Norge* was Umberto 'Airship Designer' Nobile. He later attempted to fly to the pole in a dirigible called *Italia* (he was Italian so that makes sense). The airship reached the famous pole but flew into a cocktail of bad weather that no one had ordered: fog, ice and strong wind. Ice built up on the airship, thus increasing its weight and down it went. It crashed onto the ice below and ten of its crew and tons of supplies fell out. All that Italian Mint Cake – gone! Suddenly lightened, the airship rose back up and disappeared out of sight, taking six of the crew with it, who – to follow a common pattern of epic flights – were never seen again.

Amundsen, who had retired by then, joined others in heading out on a rescue mission in a Latham 47 flying boat. Guess what? He was never seen again either. The lesson from flying over such an inhospitable place seemed to be: stay at home.

GO SOUTH

In 1929 the intrepid now-Commander Byrd and Norwegian pilot Bernt 'Pilot' Balchen set off in a Ford Tri-Motor called *Floyd Bennett.** Byrd and Balchen headed south as that's the best way to reach the South Pole. They soon reached the most southerly point on the world. Or at least flew over it. When they got there the crew spent much time flying around the pole watching the compass get very confused.

* **If that name seems familiar, it's because it was mentioned earlier, Bennett being Byrd's pilot on their 'first' flight over the North Pole. He had subsequently died, though not in a crash but from pneumonia, contracted while on a flight. Flying was really dangerous in those days.**

You need a bit of light relief on a flight that lasts almost nineteen hours. No matter how sore his tush was, the Byrd man had done it again.

GO UP

Many years ago, a climber was asked why he wanted to climb a certain mountain. He replied, 'Because it's there.' This logic was unarguable. This mountain was Everest, the highest of all the mountains. In 1933 an expedition was mounted to fly over it. Why? Because it was something to do, to ward off the inescapable empty void that haunts and confronts all human existence. And it would be something to brag about.

The project was financed by Dame 'Fanny' Houston, who saw it being a source of national pride. Unless they crashed. A British team of intrepid pilots set off, taking with them those essential items for such an endeavour: aeroplanes. They took two Westland biplanes, specially converted for the epic high flight. As the crews would take a while to get to the right altitude, they were provided with oxygen and a magazine rack with suitable periodicals. On a famous date, the two aircraft, piloted by keen young RAF officers, David 'Flight Lieutenant' McIntyre and Douglas 'Squadron Leader & 14th Duke of Hamilton' Douglas-Hamilton, took off. They were accompanied by an observer in one aircraft and a cameraperson in the other. Unfortunately the cameraperson had a problem with his oxygen supply and conked out. But on they flew into history, passing over the summit. They had done it! Lord Douglas later was involved in another famous flight, albeit less celebrated, which we must remember to include later.

GO AROUND AND AROUND

In 1930 two brothers, John 'Brother' Hunter and Kenneth 'Brother' Hunter, got into their *City of Chicago* plane and 553 hours later got out. Actually, they had gotten out while flying as their plane had a walkway to allow access to the engine, to change spark plugs and so on. This might have been the most brotherly of all flights as the aircraft was refuelled by another crewed by two other Hunter brothers: Albert and Walter. Five hundred and fifty-three hours? That's over three weeks. Can you imagine how much they wanted a bath?

Now flying around and up and down was all very well but the serious business of moving people around the planet was gaining ground.

SEVEN

CIVIL TRANSPORTATION OR I SAY, JEEVES, HAVE YOU SEEN MY BOARDING PASS?

After the First World War, people wanted to fly and could now do so without being shot at, even when flying over Manchester. With military aviation now taking a back seat, it was civil aviation that took over. Warplanes were converted to carry passengers instead of bombs, although occasionally ex-RFC pilots lapsed when flying over France and dropped a consignment of passengers onto marshalling yards outside Nantes. Hey, who said flying was without risks? Those who weren't dropped as munitions could enjoy flying in draughty, noisy, stinky, shaky, bumpy, oily and risky aircraft. Despite these drawbacks – and their appropriation for a Snow White aviation movie – civilian passenger flying grew and grew.*

The epic exploration flights had made flying exciting and glamorous. Aircraft manufacturers, being commercial operations, pondered how they could make money from this. They worked out that if they could design aircraft big enough to carry people and safely deliver them to a destination

* **Worldwide rights are available.**

of their liking, then they could receive money for this. This money would come from organisations called 'airlines', who would provide this safe passage – or at least a reasonable percentage of the time.

The very first scheduled air passenger service had been launched in Florida in 1914. Literally. The St Petersburg–Tampa Airboat Line used a Benoist flying boat to take one passenger at a time across Tampa Bay. Having one passenger isn't a hugely resilient business model. The venture lasted four months.

Soon after the First War had ended Germany and France quickly set up their airlines. Britain was slow to follow. Wouldn't these airliners frighten the horses? Germany had the first domestic airline with Deutsche Luft-Reederei (Deutsche Luft-Reederei). In 1919 it started flying from airfields in Germany to other ones also in Germany. It later merged with another airline to form one of the most famous German airlines in the world: Lufthansa (Lufthansa).

As far as we can tell, the first daily, regular,* scheduled international airline route was from London to Paris, operated by Aircraft Transport and Travel Ltd (AT&T), beginning on a date in 1919. For the discerning traveller, flying to Paris meant no visit to Dover and no ferry – two pluses right there. The aircraft used was a single-engined de Havilland DH.4a (a converted bomber type). Most of the cargo** was mail, newspapers and some grouse (they had saved up specially). The human passengers were able to sit inside – unlike the pilot – the height of luxury, don't you know. This was the start of the high-cost era of air passenger travel, with passengers being those who had lots of money. The plebs would have to watch from below.

*** Subject to baggage handler industrial disputes.**
**** Stuff carried by planes that isn't humans.**

Next to fly from the UK were aircraft operated by Handley Page Transport, which utilised its Handley Page heavy bomber fleet to become heavy people transporters. They also carried underweight people. The HPs could carry ten passengers rather than the two in a DH.4a. Unfortunately the enterprise wasn't a roaring 1920s success, with the airlines not getting enough passengers. Maybe word was spreading about how rude those Parisian waiters were. There were calls for the government to intervene and this they eventually did, setting up a committee to look into things.

Converting ex-warplanes to carry passengers was all well and good but new types of aircraft were needed. Planespotters had noted down all the current registration numbers and their pen fingers were twitchy. They wanted more, so new designs took to the air. Manufacturers in Britain stuck with making biplanes but in America, Germany and the Netherlands monoplanes came out of the factories.

German company Junkers built the first all-metal airliner, the Junkers F.13. It had one engine. One. Despite this seemingly obvious drawback, they were regarded as reliable. The Dutch company Fokker built their own machines. The first was the F.II (as in Roman two, not as in the Scottish rural greeting 'aye aye'). It was followed by the F.III, which made sense. The F.III provided interior seats for all the passengers, much more acceptable than its predecessor, which made one of the passengers (probably the one who was last to check in) sit up front and outside with the pilot.

In the 1920s manufacturers in Europe and America built airliners that looked *exactly the same*. Junkers, Fokker and Ford all made planes that had three engines: one on each wing and one on the front. They were called 'trimotors', the aviation term for 'really ugly aeroplane'. The Ford Trimotor was nicknamed 'The Tin Goose', which sounds like something the shiny chap from *The Wizard of Oz* would get arrested for. It was made of corrugated metal in order to keep the passengers awake when it was raining. Metal offered new possibilities. Apart from its structural strength and inability to give splinters, when a metal airplane caught fire it took longer to burn. This was used in Ford's marketing: 'Fly a Ford – it Won't Light up Like Kindling'.

Airliners were now connecting cities in Europe and America. In the USA passengers could fly coast to coast on flights termed 'the red eye' due to the onboard chefs' keenness on cutting up onions in the aircrafts' galley. In Europe the major airline was Germany's Lufthansa (Lufthansa), which was good news for them.

BARNSTORMING

In America, some of the war surplus airplanes had been sold off, cheap as a nickel in a dime store. They were used by 'barnstormers' – pilots who would perform amazing aerial feats in the sky, in what was also known as 'showing off' (like Charles 'Barnstormer' Lindbergh, who we mentioned earlier). They would fly under bridges, over bridges and, if positioned wrongly, *into* bridges. It was done for money and the barnstormers were competitive, always looking for the latest stunt to draw the crowds. They would fly with women playing tennis on the wings, women dancing on the wings or women dancing while playing tennis on the wings. Now this might appear fraught with danger and it was. You could pay your entrance fee and no one would fall off the wings. Hey, you pays your money and takes your choice, buddy!

As these barnstormers toured the country they would also offer locals the chance of a quick hop* in an airplane. For a

* **Did they use kangaroo petrol? #oldjokeshome**

few bucks many future pilots and astronauts got their first taste of flying this way. A certain young Neil 'First Moonwalker Oh Yeah' Armstrong was taken up in one, although it wasn't plain sailing. He had to get help getting to the cockpit, as he couldn't make it up the one small step. Ahem.

PUSHING THE ENVELOPE

People love getting mail. Nothing compares to the happiness in receiving a letter from a dear friend, a postcard from a fondly held relative or a catalogue for useful household accessories. In America after that bad old war, delivering the mail by air really took off. There were plenty of airplanes and air pilots kicking around who could continue flying. One such mail pilot was a certain Charles 'Postie' Lindbergh.* This 'air mail' service was first class, although not to the pilots' families, who didn't always see Daddy back for the weekend, or any day for that matter. Navigation systems to be developed in the future were of small comfort to those who flew into the mountains then.

NEW-FANGLED INNOVATIONS

Aircraft continued to be developed. Some bright sparks had the idea that they would go faster if they didn't have the aerodynamic qualities of a bus shelter. The Lockheed Vega was one such product of this thinking. It not only had a snazzy name, it had a snazzy shape. This smoothly shaped operator could do 135mph. And not in a death dive. In level flight.

* **Yes, this was mentioned several pages back. - Ed.**

Other new things appeared such as stressed metal fuselages (which you would be too if you were responsible for keeping a bunch of passengers in one piece). Cockpits had more instruments, not so the flight crew could entertain the passengers with this new-fangled jazz, but so that they could tell how fast they were going or what angle of bank they were in. Another one told them how quickly they were descending to the ground. This was probably the most important one.

Another new thing was the 'retractable undercarriage'. This meant the wheels that go round and round on the plane could be tucked up inside the aircraft. This made the aeroplane more streamlined and so faster. But it also gave the pilots something else to forget to do. It wasn't so bad after taking off as it just meant they flew around more slowly, but if they forgot to put them down before landing, it was a lot more messy.

GOING LARGE

At this time Britain still had its empire and it was felt an imperial airline would make it much easier to get around. This was duly done and given a simple but effective name: Imperial Airways. A couple of the types used by the airline were the de Havilland DH.66 Hercules and the Armstrong-Whitworth Argosy. They were big three-engined biplanes and looked very similar to each other, causing many arguments from the Croydon Planespotters' Group among members who couldn't remember how many tail fins each one had.

Another was the giant Handley Page HP.42, which took the Empire biscuit when it came to land-based airliners. It was the biggest airliner in the world at the time. To give some sense of its size, if your feet are 12in long, it would take 130 heel-to-toe steps to cross from one wingtip to the other. You can't test if this is true as no example of the aircraft survives; you'll just have to take our word for it. These big beauties were given individual names beginning with the letter 'H', such as *Heracles*, *Hagrid* and *Huntergatherer*. Interestingly the HP.42 was a 'sesquiplane', an aircraft that has two wings but with the lower one shorter in length. Every day is a school day with us, folks.

Despite the luxurious glamour, travelling on these big airliners wasn't all sweetness and light. On the HP.42 passengers could relax in comfortable armchairs, sip martinis and chow down on fancy meals served on fine china while sitting feet away from four 490hp Bristol Jupiter engines. The planes weren't pressurised and so couldn't climb above bad weather. As a result, there was a bit of an issue with turbulence and its companion, airsickness, which came up a few times – like the nice lunch they'd just had. No passengers were killed in HP.42s as it was difficult for them to get up to a speed fast enough to crash at. Dutch aircraft designer Anthony 'Dutch' Fokker famously said that they had a built-in headwind. What did he know?*

FILL YOUR (FLYING) BOATS

One issue faced when flying round the world concerned runways: there weren't any. Building airports was expensive and planning issues could take ages. To get round this they decided to use the water to land and take off from. After the first few aircraft sank, they reasoned that special types were needed. So, using great technological innovation, they built aircraft that could land on the water, and take off again if they wanted to. These were sensibly called 'flying boats'.

* **A lot. – Ed.**

Short Brothers (there's them brothers again) made Empire flying boats. They were known as the 'C-class' because each had a name beginning with the letter 'C', such as *Canapé*, *Castor Oil* and Cooee. One of these is actually real.

Travelling in a giant flying boat was an experience never to be forgotten – no matter how hard you tried. It was presented as the height of elegant travel: being looked after and catered for 24/7/365, overnighting in hotels in exotic destinations, having cocktails on the patio with a select group of those who could afford to travel this way. The reality was sitting in a noisy, sweltering metal tub, day after day. And you had to get up early as much of your day was spent travelling as these big babies weren't speedy. Ten days took you from the UK to Australia and your hearing would return a few months later. Passengers would arrive at their destination like their martinis: shaken *and* stirred.

In terms of the crew on board, there were the usual pilots and navigators but also a new type: the cabin crew. They were charged with filling glasses, handing out sick bags – that sort of thing. The first were known as 'cabin boys'. It may shock some, but no women were employed as cabin crew members. Eventually a woman in America called Ellen 'Woman' Church persuaded airline bosses that it would help convince nervous passengers that, if a woman was working on the plane, they must be safe. To further reassure any nervous nellies on board, these women would be nurses. It might just be us, but having medical staff on a plane as standard suggests the opposite of calm reassurance about your chances of making it through to your destination unscathed.

Over the years more and more women applied. If they had the right weight, age, size, marital status and ability to put up with male chauvinistic passengers, then the job was theirs. It is a reflection of the times that it was felt OK to employ women on how they looked, with airlines hoping that men passengers would see having attractive 'air hostesses' as a requirement before booking flights. Maybe these men should have been more attentive to the pilot's abilities with the controls.

Before this chapter ends, we should mention there were other flying boats in operation. The American company Pan American used Sikorsky S-40s. They were called 'clippers' as you had to be wary of the propellers when boarding. The S-40s were ditched for S-42s and Boeing 314s. The Boeings were big machines with dining rooms, private cabins, cinemas, water slides and driving ranges. The 314s could carry plenty of petrol, enough to make it across the sea from New York to Europe (they weren't fussy exactly where they landed) and started just in time to have the service interrupted by The Next War.

The biggest of the flying boats was built in Germany: the Dornier Do X – the 'X' standing for 'Xtremely Large'. It was powered by twelve engines, in a six-facing-forward, six-facing-backwards arrangement on top of a giant wing. The Big X could carry around a hundred passengers but not very high up as, even though it had those dozen motors, the Groß Dornier was underpowered. Several accidents weren't great for the sales brochure and only three were built.

There were other big beasts of the air around at this time and we may as well mention them here, although not all were flying boats. One was the Italian Caproni Ca.60 Capronissime (Capronissime). It's hard to describe but if you can imagine a houseboat covered with three giant venetian blinds, you're on the way. (Or you could look it up on the internet.) If you did, you'd see a nine-winged flying boat, its wings in three sets. It was described as the 'Mammoth of the Air' and it was similar in one respect: the thinking behind it was certainly 'woolly'. The Ca.60 was designed to carry a hundred passengers over the Atlantic but it didn't carry any anywhere, crashing on its second flight. An expensive way to secure firewood.

Another big beauty was the German Junkers G.38. This passenger aircraft had wings so thick, passengers could sit in them. They also gave access to the mechanics so they could tinker with the engines in flight. They must have done good work as the G.38 did carry passengers, although only two were built. One was destroyed by the RAF in the war while in Greece. What a tragedy.

Another biggie that should not be missed out, even though it was made just before this period, was the Tarrant Tabor. With its three sets of wings reaching a maximum span of 131ft, it was the biggest thing of its day. When it was rolled out, there was debate whether it needed extra ballast. Some lead was put into the nose, a bit like a plastic modeller might do with an Airfix kit.* Those of a health-and-safety aircraft design persuasion might hear the distant ringing of alarm bells. The engines were fired up and off the big tail-dragger went, across the grass. When full power was applied on the top-most-mounted engines the giant machine suddenly tipped over, burying the nose into the earth. Unfortunately that's where the crew were. It was the end of them and the Tabor's career.

EASY DC

At times, competition between rivals grabs the headlines. You only have to think of The Beatles v. The Stones, Ali v. Forman or The Borg v. Starfleet. In the 1930s two US aircraft manufacturers competed to build a successful airliner and see their competitor ground into the dust.

* **Other plastic kit manufacturers are available.**

Boeing built the 247, which could race from New York to Los Angeles in twenty hours. UA (United Airlines) ordered sixty of them, while TWA (The Worldsbest Airline) had to wait as UA had first dibs on the production line. Well, they weren't going to wait, no sirree, Bob. They asked the Douglas airplane manufacturing company to make a rival airliner to Boeing's that they could use. Douglas duly did this, making the DC-1, which was a prototype, followed by the DC-2, which could carry fourteen passengers. It was to be followed by the unsurprisingly named DC-3. This would become an iconic design. It had two wings, two engines and two pilots but it was more than this. Oh yeah.

It began life as the DST (Douglas Sleeper Transport), which might make you think they were flying railway infrastructure parts. No? Just us, then. Well, they weren't. The 'sleepers' were human passengers who could bunk down on a bunk bed. A version with seats was then built as the pillow fights were putting pilots off.

The DC-3 became a legend of the skies with thousands being built and used for lots of things like carrying cargo and people. Soldiers would jump out of them. It wasn't that they distrusted its landing capabilities but they had to, being paratroopers.

Another innovation was the Boeing 307 Stratoliner's pressurised cabin. If you have a pressure cooker, you will know it works on the principle of putting potatoes and water inside, then sealing it, before turning up the heat and then, when the potatoes are ready, it whistles. An aircraft that is pressurised is the cooker, and the potatoes are the passengers. The difference is that they are not heated up, but are safely delivered to their destination. It's not a great analogy.* The reason the cabin was pressurised was to

* **No, it's not. – Ed.**

allow the aircraft to fly higher, thus achieving better engine efficiency and allowing the pilot to avoid the bumpy turbulent weather that had turned so many previous passengers to vomit-ridden green husks. At altitudes above 10,000ft, unpressurised ~~potatoes~~ passengers would start to suffer the effects of having less oxygen in their bloodstream. Now they could sit in comfort and not worry about getting pulmonary oedema, whatever the hell that is.

SCHNEIDER DOWN

Of all the competitions held in the interwar period after the First and before the Second Terrible War, the Schneider Trophy was one of them. It was named after French businessperson Jacques Schneider Trophy and was awarded to the fastest seaplane (or flying boat) flying a certain distance. The competition began in 1913 but we all know what happened next. After that war was over, it reconvened. The competitions were generally Italy v. America v. France v. Britain, and all sorts of combinations of these battled it out in the skies of the various countries.

The races drew huge crowds, who didn't go home disappointed, as several of the planes crashed. In 1922 Britain won with the speedy performance by Henry 'Not Baird' Biard flying a Supermarine Cowardly Lion.* Biard deliberately kept the engine revolutions down in the practice sessions so as not to alert the other competitors, but when the real racing started he put 'em up, put 'em up.

In 1924 only the Americans entered a team but sportingly didn't hold the event. They would definitely have won. Or come last; depends on if you're a fuel tank half full or

* **Sea Lion. – Ed.**

half empty type of person. In 1931 France and Italy withdrew (the Americans weren't bothering to enter that year), leaving Britain as the only entrant. It sportingly decided it was best for the competition if Britain went on and won. Which it did. This was the last ever Schneider as Britain had won the previous two competitions and so got to keep the trophy. The winning aircraft was a Supermarine S-6B, powered by a 2,300hp engine. That is a lot of power. After the 'competition' another S-6B was flown at over 400mph. The designer of these machines was a certain RJ 'Spitfire' Mitchell and we'll hear about him later. <Makes note>

We'll hear about him really soon as, after years of peace, something was to come along to throw a spanner in the works of civilian aviation. A large spanner with the word 'War' on it.

EIGHT

THE SECOND WORLD WAR OR THIS IS GETTING SERIOUS NOW

The first one being such a smash, the sequel was always on the cards, and on a notable date in 1939 the Second Awful War began. Those located far away from Poland, Britain or Germany heard the news and figured it was coming their way so took appropriate actions: cancelling the milk and wondering how to develop flat feet.

The 1930s had been relatively quiet in terms of military flying and aircraft development. The RAF had carved out a new role for itself as aerial policepersons. 'Air control' was used in the Middle East and Afghanistan to 'control' the local populations from the 'air'. It is difficult to issue fixed penalty notices from the air so bombs were used instead. Luckily this solved any future problems in those territories.

Lots of military aircraft of the time were silver, which made them nice to look at, especially on a blue-sky day. The RAF was still flying biplanes, something they'd continue to do throughout the decade. They had aircraft like the Hawker Demon, the Hawker Audax, the Hawker Fury, the Hawker Hind, the Hawker Hardy, the Hawker Hart and just to show it wasn't a Hawkeropoly, the Gloster Gladiator.

These aeroplanes could be seen at places like RAF Hendon during 'air pageants' or what we would call air shows – a simple way of creating traffic jams throughout the summer. Thousands would get up early to sit in their cars, miss the start of the flying display and then complain about the price of the food. It could be a fun day out for all the family.

The RAF bigwigs were keen to show the great unwashed how their money was being spent. Audiences would be entertained by RAF pilots doing aerobatics and the like. One of these pilots was Flight Lieutenant 'Douglas' Bader, who went too close to the ground at too high a speed. He lost both legs and many thought that was the end of his flying career. They didn't know that Douglas B didn't really care what people thought and was to make a return to the light-blue uniform. But more of him later.

BOMBING ON

There were concerns by many – particularly those who lived on the ground – that aerial bombing was a Bad Thing. Books and films showed the terror that could result in bombardment. There was much anxiety and angst in these pre-Second Godawful War years. Populations were alarmed at the thought of being bombed to pieces; military air commanders were worried that anti-bombing talk would win out and they'd not get to bomb anything.

Commanders like the British Sir 'Hugh' Trenchard thought bombing alone could end wars and save all that troops-fighting-on-the-ground sort of thing, which was messy and expensive and led to awful poetry. Another bomber was the American General 'Billy' Mitchell, a passionate advocate of the power of the bomb for defence. He demonstrated this power by dropping them on old navy ships in a public demonstration. Why use ships to fight ships when aircraft could do it from the air? Some people in Japan scratched their chins and thought: why not indeed?

It became like a fad. Bombers were used by Japan against Chinese cities and Italy bombed Abyssinia (aka Ethiopia). There was also bombing in Spain in the Spanish Civil War. The war was between Spanish people, but others joined in. The Germans, keen to show they had lost none of their war-making prowess, took some aircraft and carried out what would be known nowadays as a 'pilot project'.* Observers, navigators and machine-gunners also went. Germany had been banned from developing military aircraft after the First Worldwide War but basically flicked the vees and started making its own military aircraft. Without any spoilers, it wasn't going to end well.

This bombing in Spain had a major effect on the civilian population. And by 'effect' we mean 'die'. The cities of Madrid and Guernica were heavily damaged and the only good thing to come out of this (and it was quite small relatively) was an amazing painting by the painter Pablo 'Painter' Picasso. The destruction caused many to pause. Town planners in Birmingham and Dundee looked on enviously: how could they get in on this action?

Destruction from the air was thought to be so powerful it could stop war, as the thought of all that carnage would stop others from attempting it. While there had been a policy of not spending a lot of money on military aircraft during the Depression, this changed when it looked like Nazi Germany was spending a lot of money on military aircraft (and tanks, ships, guns, mess tins, etc.). British prime minister Prime Minister 'Stanley' Baldwin said, 'The bomber will always get through a lot of money.' As many had realised, defence spending could

* **<Groan> – Ed.**

be lucrative if you could get yourself a government contract, or 'licence to print money' as they were known.

In case the deterrent didn't work, air defence systems were built up: radar, fighter planes, people pushing small bits of wood around on a map – that sort of thing. Civilians were given gas masks (most feared chemical weapons more than an attack on their human rights). Bomb shelters could be acquired and then placed in a garden – preferably your own. Many civilians who were riled up at the thought of Herr Hitler were disappointed when they weren't issued with anti-aircraft guns. They were allowed to give vent to their feelings through strongly worded letters to the papers, but it wasn't the same as firing hundreds of bullets into the air at any bomber that dared interrupt Sunday lunch.

BLITZKRIEG BOP

Adolf Hitler behaved like a real Nazi when it came to his attitude to ethnic minorities, other nations and anyone he didn't like. He built up a navy, army and air force and was going to use them, come hell or high water (for his submarines and ships, it turned into low water, as they were eventually sunk but that's getting ahead of ourselves).

Hitler employed a large (though not that jolly) fat man called Hermann Goering to be in charge of ordering large buffets and fine wines and stealing artworks. He also was in command of the Luftwaffe (Luftwaffe), deciding on what colour the planes should be painted, what they should do, where they should release their incendiary devices, etc. Hermann's Luftwaffe was to play a key role in the war about to break out.

The Germans took over various territories and then invaded Poland in 1939. Britain and France took exception to this and declared it was time for a war. When it was declared, the air raid sirens went off in Britain but there were no air raids: the German bombers weren't that fast. They were fast enough, though, and their fighters were faster. The Messerschmitt Me 109 was one of the great wartime fighter aircraft. It was small, nimble, single-engined and equipped with machine guns and cannon. These cannon fired fewer bullets but the ones they did fire were bigger. When they hit, you knew about it. In a fiendishly clever scheme, one of the cannon was *inside* the engine compartment and fired through the propeller's nose cone, making it much easier for the less-than-great pilots to hit something. One of the main contentions about this aircraft is: should we call it the Bf 109 or the Me 109? This has occupied aviation enthusiasts and historians for many decades and, like the use of its or it's, it's caused many a heated argument.

The Me/Bf 109 was designed by the great German fighter designer Willy 'No Come Back Again' Messerschmitt. He also designed the twin-engined Me 110 but nobody's perfect. It did become a good night fighter but so did the British Boulton Paul Defiant – the only wartime fighter not to have any guns that fired forward. It had a massive four-gun, 1-ton gun turret mounted behind the pilot. When first used, it caught out German pilots. But the pilots that survived weren't daft and worked out if they attacked from the front, they could experience whatever the German for 'make hay while the sun shines' is.

In 1940 the Luftwaffe were heavily involved in the invasion of Western Europe. The army were doing it, so they didn't want to miss out. One of their chief weapons was the Junkers Ju-87 Stuka. It was a dive-bomber in that the pilot would climb to a reasonable height, select his target, then point the nose right down at it. The aircraft would then dive towards it. The German designers fiendishly put a pair of sirens in the undercarriage legs that emitted a very annoying whining noise during a dive,* intended to put even more fear into the poor b*&$%rds down on the ground.

The Stukas formed part of the 'Blitzkrieg' (Blitzkrieg) battle plan where tanks, artillery, troops, aircraft and horses would all drive the advance forward in a fast way to deny the enemy time to marshal a half-decent defence. It worked, as the Germans advanced through the Netherlands, the Belgium and into the France. The French and British attempted to stop them but it wasn't easy as the Germans seemed to be better at war fighting. The RAF high command had sent aircraft to help out but they were more on the chocolate fireguard spectrum of effective combat aircraft. The Boulton Paul Defiant we have mentioned. Another was the Fairey Battle, a bomber able to carry a bombload equivalent in weight to a large box of crisps. The Bristol Blenheim wasn't much better. Many brave crews were killed in these machines in a scenario that in more litigious times might be regarded as corporate manslaughter.

As the Germans pushed on, there was pressure on the new British prime minister Prime Minister 'Winston' Churchill to send Spitfires to France. This was resisted by his Fighter Command commander 'Hugh' Dowding, who said that he could get to France. The Spitfires were too pretty to be turned into wrecks over there. They were also fast, nimble and expensive.

One good type that was used in France was the Hawker Hurricane. The Hurricane has long suffered as being the Syd Little of the Little and Large combo when up against the gorgeous-looking Spitfire, but this is unfair. Without Syd Little, Eddie Large would have just been a man standing saying punchlines that made no sense without the set-ups. It

* **Imagine Joe Pasquale. But about to drop a large bomb.**

wouldn't be the same and the Spitfire on its own wouldn't have managed without the Hurricane. Why? Well, the Hurricane was sturdier and could manage better at taking off and landing from rougher airstrips. The Spitfire had a narrow-track undercarriage, which meant its main wheels were closer together and therefore it was a bit more wobbly going over uneven ground. The Hurricane could take punishment a bit better and was more easily repaired. Crucially, there were more of them when it came to the next big battle of Britain's war: the Battle of Britain.

THE BATTLE OVER BRITAIN (LITERALLY)

Never in the field of aviation history have so many books been written for so few readers by so many writers. There are hundreds of books about the Battle of Britain. But what was it? It was a battle that officially lasted for four months in the summer and autumn of 1940. German bombers, escorted by fighters, flew over the English Channel to drop bombs on airfields, factories, docks, cows, etc. The RAF, using clever radar equipment, knew where they were and so could send up its fighters to shoot at the German aircraft. Both sides lost aircrew and planes but after a while the Germans figured the British weren't going to surrender any time soon and so resorted to bombing cities at night, which made shooting at them much harder. Although short, it was much remembered as:

1. It was the first aerial-only battle in war.
2. It took place over Britain.
3. A great movie was made called *Battle of Britain*.
4. There are hundreds of books on it.

THE WAR IN THE AIR AT SEA

There were two major forms of the air war at sea: by planes who lived at sea and those who lived on the land. In the Pacific, aircraft carriers featured a lot. The American navy had aircraft like the Hellcat, Helldiver and Avenger, while the Japanese had the Betty, Val and Kate, which sound more like ladies who lunch rather than a determined attack fleet. But this they were. The Allies had given the Mitsubishis, Aichis and Nakajimas friendly sounding Anglicised names for some reason. These ladies got up to some mischief, though, sinking warships and aircraft carriers.

It had all kicked off in 1941 when, on a fateful date, hundreds of Japanese aircraft took off from aircraft carriers and headed to their target: the American navy base at Pearl Harbor, in Hawaii. When they got there the Japanese commander radioed a message that would later be used as the title of a film: *On Golden Pond*. No! It was *'Tora! Tora! Tora!'*, which is Japanese for something, something, something. Dive bombers and torpedo bombers rained bombs and torpedoes down and along into the ships of the US fleet, sinking or putting holes into eighteen of them. The Americans were stunned. They responded by saying, 'Let's talk, there's no need for further violence. Can't we end this amicably?' Of course they didn't! It was War.

The Americans responded with a daring and epic flight. To surprise and confuse the enemy they sent land-based planes to be flown at sea. A squadron of B-25 Mitchell bombers was daringly launched from an aircraft carrier towards Japan's capital, Tokyo. It didn't turn out super great as some of the American airplanes flew on to China and then in retaliation the Japanese killed thousands of the Chinese people, who hadn't really done anything. War is hell and unfair.

As the war continued, the might of American industry was able to produce more and more aircraft, aircraft carriers, guns, helmets and those things you put around the bottom of your uniform trousers to stop insects getting in.* The Japanese couldn't compete, even when they upped the overtime rates in the aircraft carrier factories. They turned to desperate measures. Pilots were selected for a 'special' mission. They thought there was something strange when landing practice wasn't part of the training syllabus. When it was explained what they were to do, many did a big gulp and quietly edged towards the exit doors. But with loyalty to the Emperor high, they figured, how bad can it get? Around 2,000 kamikaze pilots found it was pretty bad.

We should have mentioned it earlier, but the Japanese attack on Pearl Harbor was inspired by a British raid. The Royal Navy's Fleet Air Arm were still flying biplanes: the Fairey Swordfish to be exact. These 'Stringbags', as they were affectionately nicknamed (because they were great for carrying onions), launched torpedoes at the Italian fleet at Taranto. They also damaged the German warship *Bismarck* in the film *We Sunk the Bismarck*.

Despite being really old-fashioned, Swordfishes were flown throughout the war. Other biplanes kept on past their fly-by date included the British Gloster Gladiator and the Italian Fiat CR.42. Biplanes are like that old jumper you still wear. Yes, it's seen better days, yes, your partner hates it, but you can't bear to part with it. Of course, jumpers couldn't get you shot down by a much faster jumper and that's where an admittedly poor analogy ends.

For aircraft living on land the chief role was looking for submarines. The Germans had lots of submarines – called 'U'-boats because they went 'Under' other boats and fired

* **Puttees. – Ed.**

torpedoes at them. The German submarines were very good at firing torpedoes and aimed them at ships carrying essential supplies to Britain: fuel, bullets, toilet paper and pasta. These U-boats were a complete pain in the Atlantic. Something had to be done and so long-range aircraft were acquired that could fly over the sea to escort the ships and, if any submarine popped up, they would drop explosivey stuff on them.

Crews on PBY Catalinas, Short Sunderlands and B-24 Liberators were helped by clever electronic radar equipment and looking out the window. It could be very boring, flying hour after hour over a grey sea, wondering if there were any of those nice custard creams left in the plane's galley. But whether there were biscuits or not, these aircraft helped to win the Battle of the Atlantic, as it was called. It didn't get a film as who wants to watch a movie of people eating biscuits?

BOMBED OUT

After a couple of defeats in 1940, and with not much else to do, Britain started really bombing Germany. Normally when you're at a loose end you might tidy up the garage or freshen up the paint in the hall, not unleash a storm of aerial destruction. But there was a war on. The boss of RAF Bomber Command was a man called Arthur 'Bomb Them' Harris. He famously said, 'A lot of people say that bombing can't win a war. I'm not going to say that as I'd be out of a job. Duh.'

It was one of the paradoxes of the war that some (Harris) thought the German population would falter under intensive bombing, when the British population hadn't during the Blitz. The Blitz, as anyone who has read books about the bombing of Britain by the Luftwaffe will know, was a bombing campaign by the Luftwaffe on Britain. It started in 1940 and ended in 1941, with an encore later in the war. Apart from standing in the street, shaking their fists, the defenders fired anti-aircraft shells at the bombers and also hoped they'd fly into 'barrage balloons' – big balloons tethered near key targets. The Germans were crafty though and flew *above* the balloons and came at night, so making it harder to be seen. They also used clever electronic equipment to help them reach targets like Coventry, London and Clydebank. Luckily for Britain, if not the Soviet Union, the Luftwaffe switched its intentions to the Eastern Front in the middle of 1941. 'Phew,' said the high-ranking commanders, aware of just how expensive those anti-aircraft shells were.

Much effort was put into the dropping of British bombs on German people and places. At first, navigational accuracy wasn't that great and the intended target was often missed. To be fair, travelling at 200mph at night at 20,000ft, while being shot at, makes it not a simple task. Residents of neutral Switzerland shouted, 'Hey, what've we done?' as bombs intended for Frankfurt dropped into their fish ponds. Rather than wait for accuracy to improve, a new type of bombing was introduced called 'area' or 'carpet' bombing. It's a bit like making the hole in a putting green 20ft wide. How can you miss?

To get more bombs dropped, bombers got bigger. The skinny and two-engined Handley Page Hampden was replaced by the bigger and chunkier Vickers Wellington, then it got replaced by four-engined 'heavies' in the shape of the Short Stirling, Handley Page Halifax and Avro Lancaster. We must not fall into hyperbole but the Lancaster won the war. It had four Rolls-Royce Merlin engines and a crew of seven. Accuracy improved using electronic cleverness in the form of H2S radar, which was a cost-effective and reliable way of getting crews to their destinations, something its successor HS2* could only dream of. 'Pathfinders' were used. Planes flew ahead of the main bombers, dropped flares to show them where the target was and then shouted, 'No, over there!' at crews not that sharp on the uptake.

One bomber we should definitely mention here was not that big, but it was fast. The de Havilland Mosquito is familiar to anyone who's watched *633 Squadron* or the completely dissimilar *Mosquito Squadron*. The Mossie first went into action without any defensive weapons. They hadn't forgotten; it was so fast, German fighters couldn't get near. Due to its construction it was known as the 'Wood and Metal and a Bit of Rubber Wonder'. Powered by two Merlins, it proved to be a versatile design, being used as a fighter, a bomber, a reconnaissance plane and would have cut the grass if they could have fitted mower blades. Mosquitoes carried out accurate low-level raids, including one on a prison in France. Well, no one likes prison, especially those in them.

There was another raid that was very accurate, immortalised in the film *Those Damned Busters*. A squadron of RAF airpersons led by alliterative leader Guy 'Guy' Gibson flew Lancasters at very low level to the Ruhr area in Germany. When they reached some lakes that had dams at the end of them, they flew down to a height of 60ft above the cold, angry lake water and dropped a spinning bomb that spun towards the dam, hit it, descended to a certain depth, then exploded. This was done several times and some of the dams were breached. Despite this example of bravery, skill, technical accomplishment and determination, much of the present-day debate centres on the name of the squadron commander's dog: ******.

Bombing was a dangerous occupation for all concerned but each bomber had its own defences, with gunners part

* **Oo, topical satire. – Ed.**

of the crew. One sat at the very back at the rear end of the fuselage, not because the rest of the crew didn't like them, or they smelled, but because that's where a set of machine guns was fitted. These were the 'tail-end Charlies', even though most of them weren't called Charlie. They had names like Fred, Gilbert, Harry (Britain) or Chuck, Brad and Hiram (USA). The American Boeing B-17 bomber had open windows for the gunners to fire through. Everyone likes a bit of fresh air, but this was a bit much at 20,000ft.

The bomber gunners were there to shoot at enemy fighters before they got shot at. The enemy fighters, even when close by, were much smaller than the big bombers. They were fast too and tricky to see, let alone pepper with bullets. The big bomber formations made life easier for the fighters in flying nice and level. The idea was to give each other protection but it also made for a giant target in the sky. The Americans flew during the day and their engines produced lovely long white contrails, so even the doziest German pilot could spot them many miles off.

Escort fighters were brought in to help out, and one in particular was to really help out. The North American P-51 Mustang was designed and built in just over a hundred days. Nowadays it would take that time for the designers to book a meeting room on the online room booking system. When these North Americans fitted the fabled Rolls-Royce Merlin engine it produced one of the top fighter planes of the war. These Mustangs could escort bombers all the way to Germany – and back – which was important as the bomber crews could get a bit lonely. Pilots liked to personalise their machines and gave them names like 'Grim Reaper', 'Double Trouble' and 'Conscientious Objector'.

Despite all this, life expectancy wasn't high. Being a bomber crew member wasn't a great way to see a return on your pension payments. Half of all RAF bomber crew members were killed, wounded or ended up digging tunnels in prisoner-of-war camps.

Another job that wasn't super beneficial to longevity was the shooting down of unpersonned cruise missiles. Otherwise known as V-1 'Doodly Bugs', these little beggars were fired off from Occupied Europe towards Britain later in the war. They flew quite low and fairly slowly, so fighters could easily chase and shoot at them. Now there is an immediate health-and-safety issue with shooting at things carrying explosives. Pilots developed a clever technique of staying in their huts reading magazines and letting the anti-V-1 artillery take a turn.

The V-1 was followed numerically and chronologically by the V-2, which was a different kettle of high-explosive fish all together. The V-2 was a ballistic missile, fired up very high, allowing it to come down at a very high speed. They were impossible to shoot down. The only defence was to attack the launch sites, if you could find them, and the Germans weren't handing out any maps. The Nazi brain behind them, Wernher 'von' Braun, later became an American who designed the Saturn V moon rocket, showing that if you're skilled enough, Nazi bygones can indeed be let as bygones.

HESS GOES TO SCOTLAND

A strange incident took place in 1941 that has, to the delight of conspiracy theorists, never been properly explained. On a famous evening that year, a lone aircraft was tracked coming from Germany. It flew over Scotland, then disappeared. The aeroplane was a Messerschmitt Me 110, normally crewed by two, but this one had only a solo occupant, none other than Rudolf 'Nazi' Hess. He was Hitler's deputy and what the hell was he doing parachuting into Scotland? That is the question that has caused so much query, questioning and quizzing since that night all those decades ago. He stated he wanted to meet a member of the British aristocracy called Douglas 'Douglas' Hamilton. But was this true? Was it perhaps Radio 1 disc jockey David 'Diddy' Hamilton? We will never know as Hess died after spending the remainder of his life in jail. Now, if he was good at building rockets, then maybe things would have been different.

UNION OF SOVIET SOCIALIST RIGHT GOOD AIRCRAFT

It is easy to forget that the Soviet Union made aircraft during the war. They were made by people called Tupolev, Ilyushin, Yakovlev, Polikarpov, Lavochkin and not all were arrested by Stalin's secret police. Some of the stand-outs were the Tupolev Tu-2 bomber, Il-2 Shturmovik ground-attack aircraft and the Yak and Lavochkin fighters. They were good and a lot of them were made. The Soviets had been invaded in 1941 by the Germans who, like a business opening shops just before a recession, had overreached themselves. After a good start (or bad, depending on where you stood) the Soviets regrouped and pushed the Germans all the way back. By 1945, the Red Star would be seen over Berlin.*

* **Belated spoiler alert.**

TRANSPORTERS

While fighters and bombers might get a lot of attention,* it would be amiss to miss out other types that played an important, if dull, role in the war. One was the transporter. Slow and bulky, they carried things like ammunition, spare blankets, those things you put over your boots, and people in the form of soldiers. One particular type of soldier transported was the paratrooper. These were a rugged breed with attributes needed for war: aggression, determination and a hard hat. They also had something missing, the something that tells normal folk not to jump out of a perfectly working aeroplane to then battle with the far more numerous enemy below. Paratroopers were usually the first at a battlefield, floating to Earth underneath a large silk parachute, thus presenting a tempting target to defending troops. At Normandy, Arnhem and the pubs of Aldershot these battle-hardened people got a chance to show if they were good at fighting.

Not all paratroopers used parachutes. What? Eh? How? Those found to be allergic to silk were taken in a glider. It's a tough call to say which mode of war transport was better. At least with a parachute you had some control over your fate. In a glider you were tightly squeezed in beside a squad of nervous other troops, in a flying machine that was not armour-plated. Gliders tended to be made of ply board and prayers. They were towed towards their landing sites by aircraft like the four-engined bombers we mentioned earlier. At a suitable moment they were left to their own devices.

* **They are cool, though, especially fighters. Having said that, during the coronavirus pandemic it wasn't Eurofighter Typhoons flying toilet rolls into the country, was it?**

There were two pilots in a glider: the main pilot and a second one whose job it was to prevent the main one from escaping. These gliders weren't too aerodynamically gifted and so came down fairly rapidly. Landing in the dark in one of these would test the mettle and underwear of anyone. The defending troops weren't keen to receive them and so flooded fields or laid out jaggy metal obstacles to prevent a smooth landing. Many gliders were lost and if you survived the landing, this was just the start. You then had to get out and fight. And we complain about lengthy team meetings.

GOING TO GROUND ATTACK

Other roles that aeroplanes did in the war included 'ground attack'. Troops on the ground were cheered by the sight of their own side's aircraft coming over at low level and blamming away at the enemy. They booed, however, if the same aircraft came blamming at *them*, as happened a few times. These incidents spawned the mother of all euphemisms: 'friendly fire'. While there is no defence, at 400mph one platoon of dug-in troops looks pretty much like another and they were wearing camouflage after all.

When they were hitting the enemy, ground-attack planes could be very effective. Some aircraft like the Hawker Typhoon, the P-47 Thunderbolt, the Il-2 and the Ju-87 were all good at shooting at and dropping things on troops or tanks. One German pilot, Hans 'Ulrich' Rudel, is thought to have destroyed 500 tanks, which is showing off really. In Normandy, once D-Day was out the way, the Allies had 'air superiority', which is a fancy way of saying they were the top dogs. Their ground-attack aeroplanes would attack the Germans at will. It led to German cars, trucks and trains – or at least their drivers – being scared to go out during the day in case they were sprayed with bullets or rockets. Nobody is paid enough for that.

ACES PART 2

The Second Global War also had its Aces and no one was more of an Ace than Erich 'Ace' Hartmann, who, as his name suggests, hailed from Russia. No! He was German, of course. Hartmann shot down 352 aircraft, all but two of which were Soviet. Did he have a grudge against them? Possibly. He died in 1993, so we'll never know. Hartmann was the hot shot who reached the highest tally of any pilot by using skill, judgement and his Messerschmitt Me/Bf 109 to good effect. He only started flying in the war in 1942, so his tally is even more remarkable. Fourteen other Luftwaffe pilots each shot down more than 200 aircraft and there were loads who shot down a hundred. To put it into context, the top British Ace was Johnnie 'Johnnie' Johnson who shot down thirty-eight aircraft in the whole war. The reason the tallies were so different was that the Luftwaffe faced inexperienced Soviet pilots, while the RAF faced the Luftwaffe.

THE BIG ONES

No mention of aviation in the Second World War can omit how it ended.

Japan had received many, many bombs courtesy of the American Boeing B-29 Superfortress. These were luxury machines for the crew, with pressurised cabins, three-course meals and a free bar. There was an issue with the engines but not all of them went on fire. More B-29s were lost to fires and accidents than Japanese defenders. Fact.

The B-29s dropped your bog-standard explosive bombs and incendiaries and had laid waste to cities like Tokyo but they were to use not-normal bombs on two others. Hiroshima then Nagasaki were the first to experience an 'atomic bomb' – a type of bomb that had more explosive power than was ever thought possible. It is not known how many on the ground died when the bombs went off, but it was a lot. The bombs and their fearful effects helped persuade the Japanese that this war was over. After six years of bombing raids, plus a three-fronted assault by the USA, USSR and UK, plus all the other Allied nations, Germany then Japan had surrendered. It was over. They'd done it!

It wasn't to be the end of bombers in war, however. Oh, no.

NINE

THE JET AGE OR THOSE EJECTOR SEATS SOUND A GOOD IDEA

The post-war period after the war is dominated by one thing: the jet. Not the shiny black stone found on the Yorkshire coast but a new type of propulsion. There is no point getting dragged into a boring technical explanation of how a jet engine works. Basically, air comes in at the front end and goes out the back end. That's enough to be getting on with. It's better than a propeller engine because it makes an aircraft go faster and louder. Faster and louder, in most things – apart from children playing the drums – is always better.

The first jet aircraft to fly was the Heinkel 178. It took off in 1939 and landed not long afterwards. The Germans had done it! They were also the first to get jet-engined machines into combat with the Arado 234 – the very first jet bomber aircraft to go into service, and the first fighter: the Messerschmitt Me 262. This was faster than anything the Allies had, by about 100mph. That is quite the difference. You know when you're doing say, 70mph, and a car overtakes on the motorway and everyone in the car says, 'They must be doing a hundred at least,' but they are probably only doing 75 or so, because it's really the relative difference that matters and a car going just 5mph faster than you appears a lot faster? Well, it's not really like that, but we couldn't think of anything similar. Basically the 262 was fast and therefore more difficult to shoot down. Luckily the Allies had an ally on their side: Adolf Hitler. Eh? Well, his desire to use this new wonder weapon as a bomber took them away from fighter defence where they were having a real impact – literally in the case of the bombers whose fuselages they were peppering with cannon fire.

Another non-propeller type the Allied bombers had to face was the Messerschmitt Me 163 Komet (Komet). The Komet was rocket-powered in order for it to quickly reach the bombers, shoot at them and then return to base. It was well named, blazing brightly through the sky, a wonder to anyone who saw it. Plus, its rocket fuel was so explosive that, if the pilot forgot to empty the tanks while landing, it could end up in a million pieces. If this wasn't bad enough, there was another issue to concentrate the mind: the fuel was highly corrosive. If the aircraft broke up on landing, the pilot wouldn't need lifting out of the wreckage so much as – **incoming gross alert** – shovelling out.

The Gloster E.28/39 looks like a tricky fraction to convert to a percentage but was actually the first British jet aircraft. Its engine was designed by Frank 'Jet' Whittle, who was later thanked for his efforts. This jet's contribution to aviation is marked by replicas being placed in two roundabouts, at the junction of the A426 and A4303 and the other at the A327 and Ively Road junction. Well worth a stop for a quick snap as long as photographers adhere to any on-road parking regulations.

Britain then made the Gloster Meteor, which did see action before the war ended, when they chased those V-1s. The

Meteor was aptly named, as many came crashing to Earth. If the air brakes were selected by the pilot before lowering the undercarriage (or was it the undercarriage before the air brakes?), then the aircraft would lose too much speed and fall to the ground. As the undercarriage was normally used close to the ground, this didn't leave the pilot much time to pray as the Meteor didn't have ejector seats. But what are ejector seats?

I WANNA BE EJECTED

Now, aviation is inherently dangerous. This is not something you'll hear on a 'Get Back Flying Most Planes Don't Crash' seminar organised by an airline, but it is true. If you can drown in a puddle, you can break your neck from a 10ft drop. And if you're travelling at 400mph and go into a puddle, things can get a whole lot worse.

Despite the flimsiness and lack of knowledge about various aspects of flight, the early pioneers didn't all die. The top speeds weren't that fast and some of the early designs were fairly stable, so if the engine cut out they could drift Earthward rather than plummet.

But if they did plummet, these pioneers had no real safety measures, save for hoping to be thrown clear of the crash. Later on, parachutes were issued. These were good as long as your squadron chums hadn't pulled the ultimate gotcha and filled your pack with anvils and underpants rather than the intended silk canopy. As planes got faster, things got trickier. As soon as you jettisoned the cockpit canopy in a fighter jet you were in a fast-moving airstream and, as soon as you let go, this could slam your little fleshy body into the tail fin.

This is when ejector seats (or 'ejection seats') were introduced. They consisted of a seat, a parachute, some snacks for the journey and, most importantly, a rocket to fire you clear of your crashing machine. Ejector seats weren't infallible. The F-104 Starfighter's fired downwards, which made for an interesting situation when close to the ground. Not all jets provided all their crew with rocket-powered seats. In the Avro Vulcan V-bomber only the two pilots got them. The other three down the back had to watch them blast off to safety, then struggle down to a small hatch to patiently decide who went first before then dropping into the airstream back towards where the four engines were, which could be on fire, hence the need to Get Out!

The first ejector seats were made by the Martin-Baker company, whose work saved thousands of lives. Not only did surviving aircrew get the rest of their lives from the company, they also got a tie as a souvenir. Not bad going at all.

THE NEED FOR SPEED

Speed is one of the key elements of flight. With a winged craft you need enough forward speed to get the air going over and under the wing so that you can make lift and so go up. We needn't go into details; that's what Ladybird books are for.

Aeroplanes got faster as engine and airframe technology and understanding improved. But it was no use going fast if you didn't know how fast you were going, otherwise how can you brag at the bar? It got competitive and, as with horses and those skinny dogs,* it was the fastest who got the glory.

It took nine years after the Wright brothers' historic and epic flight for the 100mph milestone to be reached. It took another nine for the 200 to be reached. Speeds kept going

* **Greyhounds. – Ed.**

up until, by 1931, 400mph was possible. Things then went boom – not because of the war but because of the jets that we mentioned a bit earlier.

Britain's first jet fighter showed the potential of jetness by setting a new world speed record in 1945. A Meteor given the name *Britannia*, flown by Group Captain 'HJ' Wilson, attained a whopping 606mph over Herne Bay. As if this wasn't exciting enough, the flight was flown at a maximum height of 250ft. At this speed and height it would take roughly a third of a second to reach the ground, or in this case, the Thames Estuary. Fortunately, Group Cap'n Wilson avoided this and lived to a right old age of 82.

Speed was to take another leap forwards a couple of years later with the breaking of the fabled Sound Barrier. This was Mach 1 – the speed of sound. It's around 700mph and changes depending on height, air temperature, humidity, what way the lay lines run and other stuff. A place where humidity isn't much of a problem is a desert and so American attempts to break the barrier took place above one: the Mojave Desert in California.

The Bell X-1 was a simple little craft fitted with a rocket engine. It was painted orange because it would be easier to find the pieces if it blew up. Its design of straight wings and all-moving tailplane in no way resembled the British Miles M.52, which had been mysteriously cancelled after the government of the day thought it too expensive. Being of a generous nature, they shared the research of the potentially supersonic aircraft with the USA, as part of an exchange programme. In return, Britain received a packet of magic beans.

On a famous date in 1947 that would echo down the aeronautical history decades, the X-1 was got ready for its epic flight. The pilot was Chuck 'Charles' Yeager. As he clambered into the cockpit he took a broomstick with him, as he thought it a bit messy and wanted to do some housekeeping. He had named the airplane *Glamorous Glennis*. Now, you know when you see a man carrying flowers and you think, oh, what's he done that's he's attempting to put right with that bunch of roses? Well, you wonder what Charles had done to think naming a rocket plane would fix. Just saying.

The thing with the X-1 is that it didn't take off by itself. Being lazy, it was carried aloft under a B-29 bomber up to its release altitude of 20,000ft. When it was set free, the orange machine fell towards the Earth. The airstream prevented the carrier plane's crew from hearing the pilot's yelling. Once he'd opened his eyes, Chuck started the engine. Once it was on ... whooosh – off it went, accelerating until Mach 1 was reached. The fabled Sound Barrier was broken – he'd done it! Now, we are not ones to quibble* but should the record count? Chucky did get a boost of a couple of hundred miles per hour before coming under his own steam and he did get a lift up from the ground. Just putting it out there.

One pilot who might have done it first was Geoffrey 'de' Havilland, son of the famous Geoffrey 'de' Havilland. The year before Yeager did it, Geoff was testing a de Havilland DH.108. It was a high-speed research aircraft and it was at high speed over the Thames when it broke up. De Havilland didn't survive. The famed test pilot Eric 'Winkle' Brown flew another DH.108 and figured out what had killed his predecessor. When Brown was going fast, he encountered a rather unpleasant and violent vibration. Brown was much smaller in height than de Havilland and his head didn't hit the sides of the canopy as he thought de Havilland's had. That was a dangerous aeroplane right there.

The decade after the 1940s – the 1950s – saw some of the most dramatic increases in speed. In 1956 the fabled 1,000mph was reached by one of the most beautiful-looking

* **Yes you are. – Ed.**

aircraft ever built: the Fairey Delta 2. A delta is a type of wing that is shaped like a triangle. If you type 'delta-winged aircraft' into any well-known search engine you will see what we mean. One effect of this design is you don't get tailplanes. These can be seen on lots of aircraft and are the flat bits at the back that make the plane go up or down.

On a famous date the FD.2 zoomed along at an average speed of 1,132mph. This was in level flight too. It was a giant leap for speedkind – the previous record was over 300mph less. The pilot was Peter 'Oliver' Twiss. Well done, Peter! This lovely machine wasn't just pretty; it was functional too. It had a very useful innovation in the form of a drooping nose. As the nose was very long and pointy, the pilot could see very little of front when on the ground. They risked running mechanics over or sticking the nose into the side of a petrol tanker or barrage balloon. With the hinged nose drooped, he could see! This feature was later used on another delta machine called Concorde, which we will definitely feature later.

Much of the FD.2's high-speed testing was performed in France and not long afterwards France's Dassault company produced a machine with a delta wing that wasn't a million miles away from looking Exactly The Same As The British Wing. Oh well, it wasn't the first and it wouldn't be the last time that British work was used by others.

In 1953 the Americans passed Mach 2 with the aptly named Douglas Skyrocket (it flew in the sky and was powered by a rocket). Five years later, Mach 3 (over 2,000mph – we know!) was reached, again in America. The Bell X-2 was flown by Milburn 'Mel' Apt but he was unable to enjoy his epic achievement for long as his rocket plane immediately went out of control and he and it crashed into the desert.

This wasn't going to put anyone off going fast though and another American machine was about to become the fastest of them all.

X-15 MARKS THE SPOT

It is difficult to describe the North American X-15 but it was very fast. Oh boy, was this thing fast. Faster than The Flash with tummy troubles looking for a toilet cubicle. That's how fast.

The X-15 looked mean: a long, thin, black shape with small stubby wings. At the back were rocket engines that Buck Rogers would have admired. This thing wasn't supersonic; it was hypersonic. Oh yeah. The X-15 could go at Mach 6.7 or, in old money, 4,520mph. To give an idea of how speedy this is, if you left Edinburgh and travelled to London, it would take about five minutes. The only problem would be you were now in London and away from the Athens of the North with its stunning volcanic hills and silvery waters of the Firth of Forth.

The pilot sat in a small cockpit fitted with teeny-tiny windows, intended to prevent pilots seeing just how high they were as the X-15 could go up to a height of over 67 miles, a distance equal to that between Derby and Peterborough. To go higher you needed a space rocket, which is what one of the X-15's pilots later used to get to the Moon, but more of him* later.

BERLIN NEEDS A LIFT

Following the Second Big War, Germany was occupied by the USSR, USA, UK and, despite not starting with a 'U', France. To get to the capital, Berlin, which was now in a country called 'East Germany', you had to drive along set 'corridors', i.e. prescribed routes that were to be used and not strayed from. In 1948 Joseph 'Communist' Stalin, the

* **It's Neil Armstrong.**

head of the USSR, closed off these corridors in an attempt to force West Berlin to come over to the East side. (The city was also split.)

Well, the Westberliners and the West weren't having that, oh no. They looked at these land routes that were blocked. What alternatives did they have? The navy, always keen to get involved, searched for an area of deep water in which to sail a destroyer group, but there weren't too many as Berlin is not on the coast. The air force big chiefs twirled their moustaches, scratched their chins, then tilted their peaked caps up while rubbing their foreheads. A bright spark piped up: why not fly in supplies? Why not indeed!

A lot was needed to keep the Westberliners warm and fed and so many planes were sent, transporters like C-54s, DC-3s, Avro Yorks and Short Sunderlands. And they didn't just go once. They flew over and over again, round the clock, 24/7, eight days a week. Over a quarter of a million flights were made in the end, using millions of gallons of fuel, before the USSR backed down. It was a phenomenal achievement, but some had regrets. Why hadn't they started collecting petrol vouchers at the start?

DUFF BRITISH AEROPLANES OF THE POST-WAR PERIOD

As we've seen, after the war the British aircraft industry continued making aircraft. Now there's no easy way to say this but there were some right duds. The Meteor was followed by de Havilland's Vampires and Venoms and Sea Vampires and Sea Venoms (the navy liked to have things their own way). These didn't set the heather on fire (unless they crash landed in the Scottish Highlands), although they were used in decent numbers. They weren't particularly fast and bits of them were made of wood.

At one point it was thought worth building a jet-powered flying boat. The Saunders-Cod-Roe SR.A/1 first flew in 1947. There's a reason you don't see flying-boat fighter jets nowadays and not just because the water puts out the engine's pilot light. Think of it like this, why don't you see bath tubs in Formula One? Because it's a stupid idea.

It wasn't just in fighter planes that things went wrong. In the civilian world the SR.45 Princess (also by Saunders-Cod-Roe) was a giant flying boat that arrived in the 1950s. Just as the world was looking to fly in jets, Britain made a plane suitable for a couple of decades before. Eventually, sense took over and it was quietly scrapped. It had propellers!

Another ~~white elephant~~ big beauty of the time was the Bristol Brabazon. This came about through the Brabazon Report chaired by Lord Brabazon, the stately name of the pioneer that had flown a pig decades before. This huge silver airliner – with a bigger wingspan than the future Boeing 747 – was powered by eight piston engines. Luxury was the name of the game as only a hundred passengers were to be carried across the Atlantic in great comfort. Unfortunately, this no riff-raff rule wasn't enough to attract airlines and it was quietly scrapped. The Brabazon was brabagone.

But going back to military aircraft. Like tea in a factory urn at a prescribed tea break, trouble was a-brewing in the late 1950s. It was felt by some in government that aeroplanes had had their time and that missiles were the future. They insisted that a lot of the historic individual companies like Blackburn, Bristol, Vickers-Armstrong, English Electric, de Havilland, Folland, etc., should merge with others. They also stopped the development and procurement* of aircraft that never got to fly. Fairey had proposed a fighter aircraft derived from the FD.2 but this was canned. A pity really

* **Military term for buying.**

as it would have looked 'well smart' as the kids say. The government minister behind this was called Duncan 'Plane Butcher' Sandys and his name is held in the same regard by planespotters as Richard 'Trainline Butcher' Beeching is by trainspotters.

KOREAN OPPORTUNITIES

At the very start of the 1950s, there was a feeling among some – the war-thirsty ones it has to be said – who felt there hadn't been a proper war for a few hours and felt it was about time there was another. But where? Well, Korea hadn't featured much in the recently finished global conflict, so why not have a bit of a war there? And so, a war was duly had. North Korea (communist) invaded South Korea (capitalist). The fighting took place on the ground and in the air, where the first jet conflicts took place.

Now, there's a really funny story about this.* In this post-war period, Britain decided to let the USSR have some of its new-fangled engines and so duly shipped some Rolls-Royce Nene jet engines to the home of communism. But aren't they the enemy in the Cold War? Well, yes but it was felt that by the time the Soviets used them in their own aircraft, Britain would have new ones that were better and so it wouldn't matter.

It mattered to the Americans as, in the Korean War, their F-86 Sabre fighters faced MiG** fighters using these

* **That'll be a first. – Ed.**
** **If you're wondering why MiG is spelled like this, if you look it up you'll find that two aircraft designers called Artem Mikoyan and Mikhail Gurevich got together to form a company. It's perhaps a shame they never called it 'ArG'.**

British-derived engines. Oops. You know how some things come back to bite you? These came back to shoot you. The MiG-15 was a very good fighter plane, better than the Sabre at altitude. Despite this, the Americans shot down more than they lost, so they were pleased about that. Anyway, they couldn't be too aggrieved at them Brits for giving the Soviets secrets, as in the Second World War they'd done the same. Several B-29s had flown onto USSR territory after bombing Japan and the Soviets did what any country would do: copied them right down to the rivets and made their own. These Tupolev Tu-4 **супер крепость** (Super Fortresses) gave the communists a long-range bomber force almost overnight.*

The Korean War was also noted for being the first war in Asia where helicopters were used. The Bell 47 was the main one here. (No idea what happened to the first 46.) Its distinctive bubble canopy made it look like a flying insect and proved excellent for collecting squashed bugs when in flight. As helicopters were new and fangled, no one was quite sure what to do with them. Some were used to ferry injured troops and others for cooling soldiers who were too hot by hovering over them. Other uses would need to be found to justify the procurement costs.

IT'S JUST A COLD. WAR

We mentioned the Cold War there, but what exactly was it? The Cold War was an opportunity for America (and its pals, France, UK, Italy, etc.) and the USSR to spend vast amounts on military equipment. Both sides had fought on the same side against Nazi Germany but, as soon as that war was over, they turned to confront each other, worried that the

* **If three years is almost overnight. – Ed.**

other side might invade them. To prevent this, lots of planes were needed. And tanks and ships and all that dull stuff.

BOMBING AROUND THE CLOCK

The nuclear option was available firstly to the USA, UK and USSR. This club had bombers ready to fly to their appointed targets, where they would release their ordnance, unleashing a hellish storm of carnage and destruction the likes of which the world had never seen. Or, alternatively, they would not be called upon to unleash etc. We're the-bomb-bay-is-half-full types, so we think it probably wouldn't have been the first. And so it turned out. But nobody knew that then.

Bomber crews were on standby, the military term for 'waiting'. Some did their waiting in the air. The Americans had their aircraft aloft twenty-four hours a day, even weekends, 'just in case'. If things went mushroom cloud-shaped and those pesky commies bombed their airfields, they'd be ready to head off towards *their* airfields, going 'Nah-nah-nah-nah-nah, you missed me!' Britain's nuclear bombers were similar, although it was too expensive to have them in the air all the time, and arguments broke out over who was having to fly when the test matches were on.

The Americans built the massive Convair B-36. The B-36 was really big, there's no getting around it. It was given the name Peacemaker, although it took so long to take off, it could have been the Warandpeacemaker. Boom!

But the mainstay of the American bomber fleet was the Boeing B-52 Stratofortress. It needed eight engines to get it going but they worked, as they're still going now. The B-52 is the Cliff Richard of strategic bombers: it's had many hits since the 1950s and, while some might be a bit tired of it, it shows no signs of retiring.

The 'Buff' (Bomber Unlikely to Foment Fun) was used in anger in Vietnam, Iraq and Afghanistan, and there's still time for it to be used elsewhere. Watch out, Greenland!

On the Soviet side, their arsenal included a propeller aircraft called the Tupolev Tu-95. It was given the NATO nickname 'Bear', which was better than the Antonov An-30's 'Clank' and the An-22's 'C**k'. The Tu-95 was a distinctive-looking thing with swept wings and a long and narrow fuselage. It was designed as a bomber but its main role was annoying Western air forces. With a range of over 9,000 miles, it could do a lot of annoying. A Bear could take off from northern USSR, then fly past Norway, attracting that country's escorting fighters of F-104 Starfighters or F-16s, before continuing past Scotland, where quickly scrambled* Lightnings, Phantoms, Tornadoes or Typhoons would meet them. (They've been doing this for years.) The big silver machines would then fly near to the coast of Ireland, where they would be greeted by Irish air force personnel gesticulating at them and telling them to 'feck off' – the Irish air forces' interception capability being the equivalent of a child on a beach throwing skimming stones. Eventually, the Bear would land. In Cuba. That's some way to get your log-book hours up.

Britain built V-bombers, so named because their names all started with the letter 'V': Victor, Valiant, Vulcan and Sperrin. Eh? Yes, the Short Sperrin was designed, built and flown as a back-up in case the higher-tech V-bombers didn't make the grade. These V-bombing planes might have sounded like comics but there was little in the way of entertainment from them, just potential nuclear armageddon. The Victor ended up becoming a successful airborne petrol tanker and the Valiant ended up in the scrapyard as it didn't have the structural strength to cope with low-altitude flying. The

* **Military term for 'speedy take-off'.**

Vulcans had started out in a nice all-white colour scheme but were painted green and grey to make them harder to spot and to give aircraft modellers something a bit more interesting to paint. For years they were Britain's nuclear deterrent. As nuclear war didn't break out, you could say they worked. As we said, we're the bomb-bay-full types.

Bombers have always faced the issue of being shot down, not just by fighters, as in the traditional manner, but via missiles. The bomber crews were forced to change tactics, as they now had to fly these big machines at low level, to avoid the radar beams that guided the intercepting fighters and missiles. It wasn't a lot of fun in aircraft designed to operate at high altitudes now having to be flung around valleys and hill crests. Crews ran the risk of spilling their coffee or losing a Mars bar down the back of the radio during a violent turn.

Now, with all that having to fly near to the ground and not into it – something that wasn't always achieved unfortunately – a new type of bomber had to be created. It had to be fast, nimble enough to skirt hilltops and valley floors, and powerful enough to carry a decent load of bombs a long way. In the 1960s Britain came up with the TSR-2, a design designed to go in at low level and do all the things needed. There were teething problems and, sadly for the project, it came at a time when the government was looking to save money. So, the chop was made and TSR-2 was cancelled. You can still see them at a couple of museums, where aviation enthusiasts go to weep silently at what might have been. It's like something out of *Thunderbirds*. Its long fuselage, with short, downward-tipped wings, its fine clean tail fin ... just give us a minute, we'll be fine.

Other countries were making their own bombers that could fly close to the ground, of course. The French Mirage IV, the American F-111 and the Soviet Tupolev Tu-22M were some of them. Later the American B-1 and the British–German–Italian

Tornado were also made. Several of these types had a new thing: the swing-wing. This was a clever idea, in that the 'wing' could 'swing', i.e. move if the pilot wanted to do something to break the boredom of flying around all day. There was, of course, more to it. It meant the plane could be a straight-wing aircraft for going slowly and landing, and then a swept-wing one for going fast and landing if the pilot forgot to move them back. Or forwards rather.

One of the biggest bombers of the post-war period was the XB-70 Valkyrie. It was American (of course), massive and could thunder along at three times the speed of sound at 70,000ft. It didn't make it into squadron service, but we couldn't not mention it. It was a monstrous beast of a thing. Look at it!

FIGHTING TALK

A range of fighters were produced to shoot down enemy bombers and fighters and those pesky ones with cameras snooping around the place. The Soviets came up with some classics following their British-inspired success of the MiG-15. The MiG-21 first entered service in 1959, which is a long time ago, but it is still being used by several air forces who can't make up their minds on which new one to ~~buy~~ procure. It's a delta-winged fighter (with tailplanes, strangely) that has an air intake at the front for its one jet engine. The air intake contains the radar, a clever design style that was used in the 1950s by a few aircraft manufacturers, then not used any more. The MiG was given the NATO nickname 'Fishbed', which is rubbish really as it's hardly fitting a powerful and sleek war machine and, also, fish don't use beds; they use bunks.*

* **Great. Nautical jokes. – Ed.**

The MiG-21 is one of the biggest-selling Cold War aircraft, with over 11,000 being made. If you got a ruble each time one was built, you'd have 11,000 rubles. That might be a lot of money. The 21 was a fast machine able to reach Mach 2 but, like a lot of fighters of the 1950s, its range was limited. This was solved by the carrying of fuel tanks and telling pilots not to keep the hammer down so much.

The Soviet Union had a monopoly on selling its aeronautical wares and so the countries it had invaded after the Second Huge War were forced to buy what the Soviets had. They could hardly order some British Lightnings when they were supposedly being made by the Western capitalist pig-dogs, could they? Many other countries procured MiG-21s, but it would be a very boring person who listed them.

The countries include:

Afghanistan
Algeria
Ambrosia
Angola
Arendelle
Atlantis
Azerbaijan
Bangladesh
Belarus
Bulgaria
Cambodia
China
Croatia
Cuba
Czechoslovakia
Derkaderkastan
East Germany
Eastasia

Egypt
Eritrea
Ethiopia
Freedonia
Guinea
Guinea-Bissau
India
Indonesia
Libya
Lilliput
Mali
Mordor
Mozambique
Naboo
Narnia
Neverland
North Korea
Oz
Republic of the Congo
Romania
Serbia
South Yorkshire
Sudan
Syria
Uganda
Vietnam
Vulgaria
Yemen
Zambia*

Another jet we didn't almost forget was the Sukhoi Su-15 Flagon. Looking like something a child would draw in the

* **That's plenty. – Ed.**

margins of their jotters when they should be doing algebra, the Su-15 was a sleek, purposeful design. It had a long, thin fuselage ending in an elongated nose filled with radar. Its small delta wings and all-silver appearance gave it that great retro 1960s look, although it wasn't retro then. As an interceptor, it was designed to take down high-flying bombers. In service it was able to demonstrate its capabilities by bringing down some high-flying civilian airliners. Wait, *civilian*? Yup. The most notorious was Korean Air Flight 007, a Boeing 747 mistaken for an American reconnaissance jet and duly shot down by a Su-15. All on board the airliner were killed. It wasn't to be the last civilian passenger jet shot down by military means – it's not a great look really for a country to be doing that.

The Soviets kept making fast machines. They wanted a fighter that could get up really high, go really fast and shoot down any planes they found up there. They plumped for the MiG-25 and the West gave it the name 'Foxbat', which, if real, would be a really scary animal to encounter in a dark forest on Halloween.

The Foxbat proved to be Mach 3-fast – if pilots didn't like their engines. A speed jockey could thrash them within a centimetre of their lives to get them up to that speed but it was frowned on by the higher command. And by 'frowned on' we mean: earning a trip to the salt mines.

At this time it was difficult for the West to find out much about the aeroplanes being built behind the Iron Curtain. The dream was if one was to fall into their hands. In 1976 the dreams of intelligence officers came true when pilot Viktor 'Defector' Belenko landed his Foxbat in Japan. This caused an outcry, a scandal and a broo-haha. The jet was eventually returned to its owners, only after the Americans had a *very close* look at it, so close they only had a couple of washers left over, if you know what we mean.

The Foxbat was followed by the MiG-31 Firefox, which became another Soviet aircraft to fall into the hands of the West when flown out of the USSR by American pilot Mitchell Gant.* The Mach 6 machine would have had an advantage over any American fighter but luckily the project never went into full development. Phew.

On the western side of the Iron Curtain the main makers of military fighters were, in alphabetical order, America, Britain and France. The Canadians also chipped in, making a terrific-looking aircraft called the Avro Arrow that not only looked the part but went fast and seemed ready to serve as a Cold War warplane but it was cancelled. Why? Well, we don't want to get into conspiracy theories or ideas of government incompetence and talk of fighter aircraft not being needed any more in the era of missiles, but who knows? One good thing – for America if not Avro, which lost 15,000 employees – was that some of the engineers moved south of the border to work on the Apollo space programme. Silver linings and all that.

In the years after the Korean War, the Americans seemed to produce fighters quicker than a pastrami on rye sandwich with eggs-over-easy on the side. They made what were called the Century Series, not because it would take a hundred years to work out why on Earth some of these were put into service, but because their type numbers were in the hundreds.

First was the North American F-100 Super Sabre. It had its good and bad points. Good point: it was the first American air force jet to go supersonic without having to go into a dive. Bad point: pilots could get into a 'Sabre Dance' where, on landing, a combination of the aircraft's swept wings, the pilot moving the control column about a bit too much, the application of throttle and an imminent stall could make the

* **You sure? – Ed.**

jet go up and down very close to the ground in a Very Bad Way. It cost several dance partners their lives, including one pilot whose accident was filmed. This footage helped other pilots avoid getting into similar straits. Silver lining from a very dark cloud.

Next up, the F-101 Voodoo was to escort American bombers but that wasn't progressed with, so in order to give it something to do, they made it a nuclear bomber. Pilots learned how to fly low level into enemy territory, drop their bombs and then ... er ... well, there wasn't enough fuel to come back so er, erm we were thinking you could eject and then walk back. Sure, no problem.

The F-102 was sensibly next. It was given the great name 'Delta Dagger', even if it's essentially meaningless. But then again, what is a Spitfire? This Convair jet was intended to shoot down incoming Soviet bombers. When it first flew, the F-102 couldn't break the Sound Barrier despite the sales brochure saying it could. It did get to Mach 0.98 and that's pretty good but it wasn't enough. That 0.02 per cent was crucial, especially if you've paid top dollar. The jet was redesigned using the 'area rule'.*

It was thought a good idea to have Delta Daggers carry nuclear weapons. Now, plenty of aircraft carried them in the 1950s; they were all the rage, like hula hoops and bubble gum but more destructive. The F-102s got given AIM-26 Falcons, a type of nuclear air-to-air missile. The brave pilot would fill in his will, get in his jet and then, when he saw approaching Soviet bombers, fire off this little baby, sit back and watch the fireworks.

You'd think the F-103 would be next but the fantastically named Thunderwarrior sadly never made it into the air. Thunderwarrior, though. Awesome name.

* **No idea.**

If you see a model of a F-104 Starfighter, you'd think the tailplanes had been stuck where the wings should be. But no, this is the correct design. Its wings are teeny. The Starfighter is the bumblebee of military airplanes. Nobody knows how it flies but it does, with its long, thin fuselage, those tiny wings and a big engine. This made it fast and a quick climber but not so great at the turning.

The US Air Force were going to procure a lot of them but changed its mind and so there were plenty available to be sold overseas. The 'Deal of the Century' saw several European countries snapping them up and there were no hints of any bribery by manufacturer Lockheed in connection with these deals. One of the countries was Germany, where the jet gained the nickname 'Widowmaker', and not ironically, as the Luftwaffe lost a third of its F-104s and over a hundred pilots were killed. Bad Deal of the Century.

The F-105 was the Thunderchief. Again, what on Earth is a Thunderchief? It was called the Thud, because that's the noise people's jaws made on the tarmac when they first saw one. The F-105 was a beast of a machine. It could carry more bombs than a Second World War B-17 and nip along at Mach 2 – all with one engine. The engine was also a beast. This plane was the Thunderbeast really.

The final Century jet to see service was the F-106 Delta Dart. It carried its missiles internally, not to keep them warm, but to avoid the associated drag if carried outside. At the end of their service lives, almost 200 F-106s were shot at after being converted to flying targets for missile testing. It beats being cut up in a scrap metal yard and turned into cuff links.

In Britain, the post-war fighters included the classic beauty that was the Hawker Hunter. It has been described by pilots as a pilot's aeroplane. It couldn't be a navigator's aeroplane as there was only one seat. Hunters were followed by a mixed bag. The Gloster Javelin was so solid it could have been used to plough snow. It was not built for speed. Due to a design flaw, switching on the engine's afterburners while flying under 20,000ft *reduced* thrust.

Luckily the next plane was built for speed and no mistake. The Lightning was built by English Electric, a company that had built Britain's first jet bomber, the Canberra, as well as washing machines and cookers. It certainly 'turned up the gas' when it came to making fighters as the Lightning could 'roast' anything else in the skies but wasn't so good in the 'spin cycle'. It was fast. Very fast. Mach 2 fast. Its two engines, mounted one above the other in an arrangement set to test a mechanic's patience, gave it a great thrust-to-weight ratio. This is an aeronautical engineering term to describe where the Thrust (T) must be more than the Aeroplane (A), best expressed in terms of $T(x^{2\{y2\}})/A^2$ x first number you thought of ÷ X-15 or something.

The jet was made speedy to intercept Soviet bombers – or French depending on how cordiale the entente was at the time – before they could drop the Big Ones. There were some issues with the Lightning; it wasn't all engine whines and roses. It didn't have a lot of room for fuel and so flights were short in duration, perfect for the pilot who smoked or had a weak bladder.

THE SPYPLANES WHO LOVED ME

As we saw earlier, the need to see what others have is called 'reconnaissance' by the military and 'nosiness' by civilians. With the Cold War on the go, the Americans were very keen to see what the communists had. To do this they had to fly over Soviet territory. This wasn't as easy as it might sound, as the Soviets weren't keen to just let the Americans zoom about in their airspace. To put the capitalist pig-dogs off, they installed ground-to-air missiles called SAMs (surface-to-air missiles). Guided by radar, with explosives in the front, they could ruin anyone's day. The Americans hoped if they flew high enough, the SAMs couldn't ruin Uncle Sam's day.

To this end they built an aircraft that epitomises the Cold War aerial ~~nosiness~~ reconnaissance role. It was the Lockheed U-2 and would become very well known, moving from the pages of the aviation press to the front pages of newspapers read by normal people. The U-2 was a thin, long-winged, one-engined design that looked more like a glider than a hi-tech Cold War warplane. But this it was, operating way over the front line at high altitudes. Up there the world can seem a long way away. Unfortunately for one U-2 pilot, it got a lot closer when a Soviet SAM hit his airplane. Francis 'Gary' Powers ejected, was captured and then jailed for espionage. Serious stuff and the incident caused the USA much embarrassment as they pretended the plane was looking at the weather. 'Just over our airbases?' snorted the Russian officials who had retrieved the cameras from the wreck.

Overflights of the USSR were over but in 1962 U-2s overflew Cuba and did not overlook what appeared to be nuclear missiles. This started the appropriately named Cuban Missile Crisis. The world waited agog as the Americans and Soviets squared up to each other. Would nuclear war wipe out humanity? As we're all still here,* we can say: no.

The Cold War was thought to stop any prospect of a major nuclear-argh-we're-all-going-to-die event. But did it stop other wars?

* **Written April 2022.**

TEN

MODERN WARS OR ALL YOU NEED IS LIVE PRECISION-GUIDED MUNITIONS

No. There were still plenty of wars around. One of which resulted in much pain, agony and many Hollywood movies.

WAR IN VIETNAM

After fighting a war in South East Asia in the 1950s, America's military commanders prepared for a potential war in Europe. And then fought another war in South East Asia. The Vietnam War was fought between America (capitalist) and South Vietnam (also capitalist) against North Vietnam (communist) and South Vietnam's Vietcong (also communist). You'd think the mighty USA would have it easy as they'd successfully fought the Germans, Japanese and North Koreans fairly recently, but you'd be wrong.

In the air the Americans used everything they had. Thousands of helicopters ferried combat troops in and then flew injured troops out. B-52s, Phantoms, Thunderchiefs, Skyraiders and others dropped thousands of bombs onto foliage and occasionally hit an enemy target. The Vietcong were tricky to find from 30,000ft. They were hard enough to find from ground level.

Unfortunately for them, American aircraft could be found fairly easily from the ground. They directly felt the irritation of those being bombed, expressed through the medium of missiles, artillery shells and soldiers pointing their rifles upwards and letting fly. Many planes were brought down and, when it happened, aircrew were very keen on being rescued. To fall into the hands of the people they'd just been dropping napalm on wasn't great for an enjoyable experience. Their prison was named the Hanoi Hilton, ironically.

The Americans were prevented from all-out bombing as seen in the Second Worldwide War by rules they'd created, so they were denied the chance to flatten Hanoi, the capital of North Vietnam. This annoyed those in American military uniforms, who really wanted to do it. It's just as well they didn't get their way or the conflict could 'escalate', which is a military term for 'all-out nuclear war, we're all going to die'.

American F-4 Phantom and F-8 Crusader pilots battled fear and opponents in dogfights featuring MiG-17 and MiG-21 jets. The Phantom carried eight air-to-airplane missiles and sometimes they actually blew up near to an enemy fighter. It wasn't always the missiles' fault and so a special school was set up to train pilots to dogfight better. This led to the film *Top Gun*, proving that war is indeed hell.

A SIX-DAY WAR

In 1967 one of the most accurately titled wars started in the Middle East. It began with Israeli aircraft attacking Egyptian airfields. They didn't tell them they were going to do this and so gained 'the element of surprise', which is just behind earth, wind and fire in the element hierarchy. The Israelis used French jets such as Mirage IIIs and Mystères (Super and Normal) in their attacks. They destroyed a lot of Egypt's air force and were well happy with their aerial offensive. They then turned to Syria's and Jordan's airfields. There was a ground offensive* after it and this ensured the end of any further conflict in the region.

GOING SOUTH

If you were to have said to a high-ranking British officer in the 1970s that the next war they'd be fighting would be one thousands of miles away against a country that had aircraft and warships from their own country, they might have said, 'Are you from the future?' because that's what you were. Or would be. Or had been.

The Falklands are islands very far away from the UK, about 8,000 miles worth of far away. They are not very far away from Argentina and that's really the crux of the problem. Without going into long-winded research, a major Bone of Contention was which country got to fly their flag. In the spring of 1982 the islands were invaded/retaken (depending on how you viewed things) by Argentina.

Diplomats flew into action, literally, trying to avoid a Falklands War. 'Why can't you make a flag with a different one on each side?' they suggested. But once a taskforce had sailed from the UK, a war was a-coming. Britain had two aircraft carriers able to carry Sea Harriers, which, as the name suggests, are Harriers sprayed with anti-rust guard. These weren't 'proper' aircraft carriers as some navies might see

* **Army term for 'war'.**

them,[*] as they weren't able to carry planes with afterburners and fling them off the front with a steam catapult. The Royal Navy did have such carriers until a few years before, so the American taunting really hurt. These carriers had carried Phantom fighters and Buccaneer bombers. As we'll remember, the Phantom could carry eight air-to-aeroplane missiles. The Sea Harrier had two. Arithmetic fans will have worked out that's a smaller number.

Before we lose the thread,[**] the Harrier contains one of the great British aeronautical engineering innovations. When it was decided that a military jet able to take off vertically and land the same way was needed, the clever designer types scratched chins, rubbed foreheads, etc., etc. They came up with an idea: have a normal-looking aeroplane but stick a second engine inside that could provide this upwards and downwards movement. Slight problem: once it was in the air this engine would now be dead weight. The Harrier's designers thought this was daft. They cleverly got round this problem in the way that the future F-35 designers didn't, by having four nozzles attached to a single engine. These nozzles could rotate and direct the thrust where needed. Told you it was clever. It was so clever the Americans wanted in on the nozzle-engine action and built their own Harriers. And kept using them when Britain retired its.

But back to the war. Other aircraft took part, one of which was the elderly Cold War V-bomber, the Vulcan. It was called up to drop what are called 'conventional' bombs, rather than the nuclear, 'unconventional' bombs it was set up for. Bomb aimers ran around looking down the backs of settees for instruction manuals. Museums were raided for refuelling

* **America.**

** **And the reader. – Ed.**

probes.* Mint Cake was sourced. On a notable day a Vulcan set off south on an epic flight from Ascension Island, which is in the middle of the Atlantic Ocean. The lone bomber flew on through the night, hour after hour, the crew quickly running out of things to spot for I-Spy, in their cramped cockpit mostly painted black. Their backsides weren't getting any comfier on those metal ejection seats either.

After many hours the plane dropped to low level over the cold, angry sea so that Argentinean air defence radars wouldn't detect it and spoil the surprise. Then, at the right moment the big jet rose up to a height where the bomb aimer took aim. The bombs were duly dropped and the Vulcan quickly turned north. Critics, i.e. the Royal Navy, said the whole thing was a waste of time and they could have done better and that their brother was bigger than the RAF's brother but the Cold War bomber had hit the runway target it was literally aiming for and so ya boo sucks.

Aerial fighting continued for a bit, with Sea Harriers attempting to shoot down the Argentinean Skyhawks and Mirages as they tried to sink British ships. Both had some level of success. In Britain, Phantom pilots watched the footage on TV and gnashed their teeth.

COVERT AND OUT

Stealth means secret, covert, unobserved, concealed, by deception, hidden – we'll stop there. The military application is no different. As the main way of detecting an aircraft is by radar, an attempt was made to make an aircraft 'invisible' to this radar. Much secret work was done on a hush-hush, need-to-know basis in an unacknowledged, unnamed area

* **This actually happened.**

known the world over as Area 51. Where the other Areas were, no one knew. Or if they did know, they weren't saying. Or if they didn't know but were saying, then they needed a stern talking to.

It was all kept very secret until one day a photograph was released of the world's first stealth aircraft, the Lockheed F-117 *****hawk. Experts gathered around to have a good look at this new shape in the skies. It was angular, unlike anything seen outside an origami set. The underside was flat, its swept wings joined by a very long V-shaped tail. There didn't appear to be any bombs or fuel tanks hung outside. The jet was painted jet black and, to preserve the unobservable status to the max, pilots couldn't even wear bright socks.

Although the brochure said the F-117 was stealthy and couldn't be seen, this wasn't quite the case in 1999. During the air campaign over Kosovo, one was downed by a Serbian missile. It was a bit embarrassing for the Americans but, as the French say, 'That's war.'

With enormous amounts of money coming from government budgets, American manufacturers weren't done with stealth. They upsized for the next one, the Northrop B-2 Spirit. Wait, didn't we have a B-52 and a XB-70 and now we're back at 2? Well yes, we did, but a more recent type was the B-1 and so it made a sort of sense to go with B-2. Though having said that the next one is due to be the B-21. Are they making it up as they go along? Perhaps. The American Planespotters' Association are not overly pleased with this nomenclature mucking about.

The B-2 has a shape familiar to aerophiles who recall the 'flying wing' of the 1950s. No tail fin, no tailplanes, not even a proper fuselage. It is just a great big wing (and engines, windows, seats, drinks holders, etc., of course). The B-2 looks more like a boomerang than a bomber but, while one of those can be acquired at an Australian gift shop for $20 Australian dollars, the other couldn't be had for less than $737 million American dollars. That's never been a little amount of money. $737 million gets you an airplane that can fly over any well-defended country and bomb it without being spotted on radar and fired upon. Seems a good idea, especially welcomed by the crews assigned to fly in them. As the aircraft is named Spirit, each one is specially named: *Spirit of Texas, Spirit of St Louis, Spirit of the Blitz, Spirit of Christmas Past* to give four.

They can fly for a long time. A long, long time. Over thirty hours a mission is common and one set an endurance record of forty-four hours in the air. That's almost two days and you have to wonder at what the pilots talked about during that one. Astronauts are most often asked how they went to the toilet, but as we don't know any B-2 pilots we can't ask them that. We do know they avoided sleeping during important parts of the flight by taking Naturally Attained Periods of Sleep (NAPS) but if they struggled to stay awake they could take Potions to Improve Lassitude due to Loss of Sleep (PILLS). These airmen and airwomen were certainly sorted for zzz's and whizz-bangs.

Stealth was *the* thing to have so when the next generation of single-seater combat jets were being designed, it was sure they had stealth running through them like a stick of Edinburgh Rock has 'Edinburgh Rock' through it. The American F-22 and F-35 showed that money and time were an aircraft manufacturer's dream. They were fancy and full of hush-hush, top-secret, need-to-know, mum's-the-word electronic equipment. This made them perfect targets for industrial espionage by countries we can't name for legal reasons but let's just say the Chinese Shenyang J-31 isn't a kick in the arse that far away from the F-22 in the looks department.

The F-35 came in three different flavours, some of which have a jet engine and a 'lift fan'. 'What's this?' you might ask. Well, the jet at the back makes it go along like a conventional jet but it can then turn downwards and, with the 'lift fan', it makes the plane go up and down. When the jet is going along like a normal jet, this 'lift fan' is what is known as 'dead weight'. You know when you go on holiday and pack big bath towels, but the Airbnb has towels? It's a bit like that but much more expensive, even if you've stayed in Edinburgh during the Festival.

TOP GUNS

Anyone who was around in the 1980s will remember the song 'Take My Breath Away' by pop band Berlin. There is a real chance just by reading this it will become your earworm for a few days. You're welcome. The song had featured in the box-office smash that is *Top Gun*, a film about American navy pilots with a fondness for flying fast, playing volleyball and over-acting. It starred Tom 'Cruise' Ship, who played a maverick pilot who bent the rules and played hard and fast with the regulations and only killed one of his weapons systems officers through his dangerous antics.

To aviation enthusiasts the real star was the Grumman F-14 Tomcat, at the time the US Navy's and Iranian Air Force's main fighter jet. With two wings, two tails, two engines, two crew and too little power in the first versions, the Tomcat showed what $30 million got you. It carried Phoenix missiles, which could travel 100 miles before bringing down a hostile aircraft. That is a long way and you had to be very sure what you were shooting at when you pressed 'fire'.

The Tomcat's cockpit contained some new-fangled innovations. One was the HUD (Heads Up, Dummy) to help the pilot find out how fast he was going and where he was without having to ask for directions. He also had the HOTASS system, which, although sounding like something experienced after a dodgy curry, actually stood for 'Hands On Throttle And Something System'. It was a way of having lots of buttons and switches that the pilot had to remember how to use. This system was to help with aircraft safety, so the pilot was not looking inside the cockpit all the time trying to find a decent radio station to listen to.

The F-14 was the first of the 'Teen Series' that began to be paid for in the 1970s.* This wasn't a line-up of mumbling moody aircraft covered in spots, but American fighters numbered after thirteen. Others were the F-15 Eagle, F-18 Hornet and the F-16, which was given the super-aggressive name 'Fighting Falcon'. It was also called 'Viper', although when they didn't start on a cold morning, they were given other names less suited to a family audience. These jets were very capable, hi-tech machines that could give good displays at air shows, which is what really matters.

In the 1980s new types appeared. The French Mirage 2000 was a gorgeous delta-winged beauty that epitomised style, elegance and that 'I don't know what' that is so typically French. The Soviet Union was still making planes that struck fear into the hearts of planespotters, worried they'd never get a single registration number. The Sukhoi Su-27 Flanker and MiG-29 Fulcrum were as good at the twisty-turny flying as any Western aircraft and in the case of Britain's chief fighter, the Panavia Tornado F.3, more so. These two communist jets stunned planespotters when they appeared at air shows in the West. They performed 'tail-slides', 'Cobra' pitch-ups and 'crashes'. One such crash featured famed test pilot Anatoly 'Test Pilot' Kvochur, who saw his MiG head Earthward after

* **And the final payment will be any day soon.**

losing engine power during a slow pass. He was so close to the ground his parachute was helpfully inflated by the explosion as his jet impacted Mother Earth. What's Russian for 'a close shave'?*

WAR AGAIN

The Cold War ended after the Berlin Wall was reduced to rubble – literally – in 1989. The decades-old enmity between the dour communist Soviet Union and the fun capitalist West was over. People from the West could now visit their former opponents' country and buy those big furry hats or KGB thumbscrews as souvenirs. Some also came back with bits of the Wall, broken into commercially viable fragments. Those communists sure learned capitalist ways quickly.

Western military commanders were concerned. Would they never get to use all their hi-tech equipment? Luckily a solution to their worries was just around the corner, if the corner was the Straits of Hormuz. When Kuwait was invaded by Saddam 'Badman' Hussain in 1990, Americans ran to their atlases, trying to find where the hell Queuewait was. Once they found it, their interest fizzled out. They had plenty of hot places in America; who wanted another? Then someone who was reading a gazetteer piped up that it had lots of oil.

Aircraft, tanks, soldiers and pizza-vending machines were shipped out within hours. Other countries joined in. Britain, which had not been able to drop bombs or fire missiles since the Falklands, sent Jaguars, Tornadoes, Buccaneers, Nimrods, Victors, Herculeses, Tristars and some helicopters. Many of these were painted in special 'desert' camouflage,

* **удачный побег. - Ed.**

although the Americans, who had lots more aircraft there, didn't bother. Painting planes would eat into pizza-eating time, bud.

The plan was to bomb the Iraqi troops for a while and hope they'd surrender. If that didn't work, then the non-Iraqi forces would invade Kuwait and drive the original invaders out of the country. Hundreds of thousands of troops were sent to prepare for the invasion, and looking at the size of Kuwait you wondered how they were all going to fit in. Would they need bunk beds?

The bombing campaign lasted a month, giving the Coalition's* aircraft plenty of opportunity to drop those bombs. The Americans used their new Stealth fighter (actually a bomber) and the British called up the 1960s-era Blackburn Buccaneer (always a bomber). The Buccaneer was used to fire lasers at targets. They weren't the death-ray lasers that Dan Dare fired but weak-water ones that could tell a 'precision-guided munition'** where the target was. These bombs were dropped by Tornadoes, which had begun the war by flying very low, but that is a quick way to run out of crew underpants, so they started flying a bit higher. In the face of overpowering air power, the Iraqi Air Force gathered up its courage and belongings and flew to Iran to sit out the war. Eventually this 'Gulf War' was over, but it was left hanging, ready for a sequel.

* **The lyrical name for the ... er ... coalition of countries taking part.**

** **Military term for 'fancy bomb'.**

ELEVEN

CIVILIAN AVIATION OR WHY DON'T WE GET PARACHUTES?

Not everyone wants to fly a warplane around. Many prefer to pilot unarmed aircraft, even those flying into Doncaster Sheffield. This is called 'civil aviation' and includes airliners – how most of us experience air rage. Airliners have opened up the world. Without them we might not travel to places like Ibiza, Bali or Bristol. It's as easy to get to Malaysia as it is to get to Maidstone – maybe easier with the cross-London infrastructure being a nightmare during the week these days. I mean, what are they thinking with all these emission zones I ask you.

PROPS DUE

After the Second Large-Scale War, air travel expanded as it was now safe to go to countries and not get shot down. Passenger-carrying aircraft were mostly piston-engined propeller types as that's what was available. From America came the Constellation, the DC-4, the DC-6 and the DC-7. Piston engines allowed engineers to trot out the old joke about being late because their 'piston broke' and the other person assuming they were talking about being intoxicated and without any money. It was a lot funnier when Kenny Everett told it. Piston-engined airplanes were reasonably reliable though and carried passengers across the Atlantic and America – and most arrived safely. This was an important element in encouraging air travel.

Turboprops then came into usage. These engines used a 'turbo' to turn the 'prop' and powered the Soviet Tupolev Tu-114 and the British Vickers Viscount – the very first turboprop airliner. Pop music fans will be familiar with the sound of the Viscount's engines, as they featured in Fab Four pop band The Beatles' song 'Let's Return to the Union of Soviet Socialist Republics'.

COMET THE HOUR

The first jet airliner was British: the de Havilland Comet. The Comet was graceful, curved in all the right places and with a balanced design few aircraft manage to emulate. Undoubtedly it was a lovely looking thing. Unfortunately, it was also a crashy type of thing. Several examples came down too quickly and effectively ended Britain's lead in making jet airliners.

To find out what happened, engineers put one in a big water tank and when it burst they had found out it wasn't strong enough to go in a big water tank. It also wasn't strong enough to fly in the skies: bits of the fuselage were cracking up (not with laughter) with metal fatigue. Not a good thing. In fact, a Really Bad Thing. As well as those lost in the crashes – for which it couldn't get any worse – for de Havilland it was a PR disaster, worse than when that man who owned jewellery shops said the jewellery in his shops was rubbish.

Work was done and the Comet was fixed. The issue with the Comets had been predicted by a man known as Nevil 'Para' Shute. Shute had worked for de Havilland as an engineer but left to write books. One of these was about a British airliner that broke up in the air due to metal fatigue. Which is exactly what happened in real life! Spooky or clever? We just don't know as Shute died in 1960 and we don't know any mediums.

You know that thing about clouds and silver linings? American manufacturers were able to make aviational sales hay while the sun was shining on them and not on British ones. As keen-eyed readers will know, Boeing had made bombers such as the B-17 Superduperfortress and B-29 Evenmoresuperduperfortress. They had four engines, could fly a long way and carried a lethal payload, so after the war Boeing built jets that did the same, although their payload was much less destructive: tourists. Well, possibly. Have you seen Venice lately?

The Boeing 707 was their first jet airliner. It was pronounced 'seven-oh-seven', with the 'oh' standing for 'Oh my god, you expect us to sit in that for ten hours?' It wasn't that roomy inside, but it was fast: 707s could zing along at 600mph. The main selling point for passengers was: it wasn't a Comet, so people were satisfied with that.

The Comet was worked on and flew successfully for several years. The French built something that was almost as pretty: the Caravelle. It sounds like a smooth, sumptuous and sweet confectionery, and while that can never apply to a large collection of metal parts, this French airliner was pretty delicious. It was similar to the Comet – the British had kindly helped the company making them, continuing the great British tradition of assisting the competition – but had an interesting configuration where the engines were at the back. This was to be used by many other airliners over the years such as the BAC 111 (pronounced 'One Eleven' not 'One Hundred and Eleven' or 'One One One' or 'Eleven One' – which sounds like a San Marino score line), the Trident, the VC10, the Douglas DC-9 and a few Soviet ones too. There are sound reasons for this: literally – those in the expensive seats at the front are further away from the noisy engines.

THERE IS A LIGHT AIRCRAFT AND IT NEVER GOES OUT

If you're a fan of YouTube (or any other video-sharing website), there's a chance you'll have seen footage of a plane landing on a road. The aircraft will be what is termed a 'light aircraft', not because it's painted in light colours but because of its weight. Now you might get scoffed at, or worse, if you said this to a driver crawling out of their wrecked motor from under your airplane, but it's true.

In the 1950s light aircraft took off, due to the popularity of aviation, people having more money and the awfulness of light entertainment on offer. A plane owner could strap the luggage in the seats, stuff the family in the baggage compartment and fly off into the wild blue yonder for the weekend. One plane was the king and queen of the light aircraft world: the Cessna 172. More have been built than any other airplane: over 44,000 of them. If you can imagine every citizen of Pontefract having a Cessna, that's the amount of 172s there are. Any planespotter deciding to go after their registrations is going to need a lot of notebooks.

They are still being made, decades after they started. There's an old aviation manufacturers' saying: 'If it ain't broke, don't stop making them cash cows.' People just love the tricycle-undercarriaged, high-winged, stable-in-the-stall machine. One such pilot was German teenager Mathias 'Cessna' Rust. On a date in 1987, young Rust ignored the advice of the Village People and Went East, guiding his light aircraft through the Iron Curtain. Despite being intercepted by Soviet fighters, Rusty made it all the way to Moscow and, in front of startled Muscovites, brought his plane in for a landing. What on Earth made him risk being blown to bits by a missile or aircraft cannon shell? Turned out Matty was on a mission: a mission for peace. He was certainly given a 'piece' of the judge's mind for being so daft and was sent to jail. Rust was released after a time and became a financial analyst for a bank, which is something Gandhi and Nelson Mandela would have done if they'd thought of earning a lot of money instead of selfishly advocating freedom for their peoples.

CARPLANE DIEM

It has long been a dream of many for a car that can convert into being a plane. Motorists stuck in traffic can easily drift off, imagining floating up into the less-congested sky. Of course, if all the drivers got carplanes, then the sky would become just as congested and the flying motors would need to be fitted with machine guns or air-to-car missiles to help clear the way for those who just cannot sit at traffic lights for a second longer than necessary when someone doesn't move off rapidly enough. Is it so hard to keep an eye on the lights?

It's a tough design brief, as adding wings etc. adds weight. It then needs a lot of power to get off the ground. On the ground you've got all these rudders and tailplanes sticking out all over the place, making it tricky to negotiate the tight parking spaces at your local supermarket. As we know from bashed door panels, many drivers struggle now. Stick wings on an SUV and you've got Car Park Gridlock, which sounds like a rubbish console game.

Currently there are no carplanes available to the mass market, with many preferring to wait on the arrival of low-cost jet packs. We hope they are not holding their breath, as we're waiting on food in a pill form. And are very hungry.

CARGO NOT PANTS

Cargo aircraft fly cargo around. Anyone who has seen the Tom Hanks fishing movie *Cast Away* will have seen cargo planes in action.* For years they were unsung, unloved and unwritten down except for by a very niche set of planespotters, who would get up very early to catch that mail plane heading off at the crack of dawn. Then the pandemic came and stuff like masks, medicines and jigsaws needed to be sent very quickly around the world. These cargo planes had the skies to themselves as passenger airliners were grounded, what with everyone being told to stay at home.

The biggest of them all is the Antonov An-225. It is to aviation what King Kong is to monkeys. It can carry a great deal of anything in its cavern-like interior. The 225 was built to carry the Soviet Space Shuttle rip-off but after that was ditched it was made available for hire to carry really big things around the world. So, if you have helicopters, tanks or bits of bridges needing moved – give them a call on 0202 BIG PLANE.**

* **We are legally obliged to state that not all crash and deposit their survivors on desert islands.**

** **Sadly, amongst the more important human losses of the 2022 invasion of Ukraine, this big giant was destroyed.**

PLACE OF THE CONCORDE

Speaking of British–French co-operation, in 1962 Britain and France signed a historic agreement. They were going to build an aircraft together. Not an Airfix kit but a 1:1 scale, full-size, actual airliner. It was to be called Concorde – the 'e' standing for 'expensive'. (It needed a lot of glue.) Many thought it a white elephant, but it was more of a white cheetah: skinny, very fast with a love for chasing wildebeest. Concorde was so fast it left its own cares, engine noise and luggage far behind.

Few aircraft would grace the skies or balance sheets with such breathtaking figures. It could fly at 1,400mph, at 60,000ft, at over £20 million a pop. It was classy: all seats were in first class, a great way of charging full whack for each well-heeled bottom that sat down in it. There were no riff-raff on Concorde. The A-list of the celebrity world travelled on the big white bird: the Queen, Sean Connery, Mick Jagger, Paul McCartney. And Piers Morgan flew on it too.

Concorde was designed to travel across the Atlantic but, when it did, it ran into trouble. Showing exquisite irony, Americans objected to it being too noisy. New Yorkers too! They said they didn't want it coming in and showing them that they didn't have anything as remotely fast or cool. The Americans tried to make their own 'SST' – Supremely expenSive Transportplane but they only managed a wooden mock-up, which also used up a lot of glue.

The Soviets also made their own SST (Supreme Soviet Turkey) called the Tupolev Tu-144, which looked just a little bit like Concorde and actually flew first. It crashed at the Paris Air Show in 1973. It's never a great thing to crash your plane while trying to show it off. Another crashed in 1978.

Concorde had a reasonably safe flying career (yes, a few bits did fall off, but nothing major). That was until 2000, when an Air France Concorde crashed shortly after taking off. The other planes were grounded and changes made to prevent this happening again. It returned to flight but, like that incident in the Bible, the writing was on the wall and Concorde's days were numbered: 1,186 in fact. In 2003 it flew for the last time. The Heathrow Planespotters' Group wrote down 'Concorde', the ink smeared by tears. Only twenty Concordes had been built and those surviving were packed off to museums or airfields to sit out their days, dreaming of when they were the best thing in the skies. Sniff, sniff. Lot of dust in the air today.

WIDE BODIES

In the 1970s airliners that weren't Concorde were all about having a wide-body configuration and using high-bypass turbofans. No one knew what these were, but they sounded good. Who wanted a thin-bodied airliner or a low-bypass turbofan? They didn't sound good at all.

A quick look online and you will find that 'wide body' referred to the type of fuselage rather than a type of passenger. (Although you could do a chapter on the increasing amount of person that can be seen trying to squeeze into a standard airliner seat, but for space and taste reasons, we won't.) Airliners traditionally had one aisle, with seats either side, like in a bus or a crematorium. The wide body had two aisles, and three columns of seats. Like in a wide-body airliner.

The first of the breed had several names: the 'Queen of the Skies', the 'Jumbo Jet' and the 'Boeing 747'. It weighed over 300 tons and had sixteen wheels. It could carry over 600 passengers. It had many toilets.

The 747 ushered in a new era, as in the 1970s mass air travel really became a thing. People who once thought big metal birds in the sky were not for them could now afford to sit in an uncomfortable seat for hours, just like the toffs and celebrities. And, despite having to use the on-board toilets, they came back and flew again.

Some mourned the loss of the great days of air travel, of silvery Empire Biscuit Flying Boats and silky smooth Comets. But none of those could get you to a two-week all-inclusive in Alicante for under £100. Passengers enjoyed Mediterranean summer sun destinations but, when they got bored of trying to find a café doing a proper full English breakfast, began looking further afield. Asia became more popular, allowing holidaymakers to enjoy the company of sleep-deprived parents and their colicky children for much longer. This burgeoning long-distance travel came at another cost: returning holidaymakers could now bore the arse off their friends with photos of Bali and Goa rather than Majorca or the Greek islands.

As the world was getting smaller, planes were getting bigger. The wide-bodied McDonnell Douglas DC-10 (Ten) and the Lockheed L-1011 (Ten Eleven) TriStar started edging out the thinner 707 (Seven-Oh-Seven) and DC-8 (Eight). These newbies had an engine arrangement that combined the traditional 'engines under the wings' beloved of designers and the mechanics that had to service them, with the 'engine in the tail fin' that looked strange and was a pain in the access ladder. The DC-10 was first into service, but its years weren't marked with images of cruising through cloudscapes. Instead, a series of 'incidents'* ensured the jet was more associated with footage of ambulances and air accident investigators standing in charred fields. The plane earned a reputation as unsafe and was grounded at one point. Several cargo doors fell off and an engine departed during

* **Crashes.**

one plane's take-off. 'So what?' said the company's bullish PR officer. 'It's got another two. Stop making such a fuss.'

The TriStar came second. It was technically more advanced and had an 'autoland' system that could 'land' the plane 'automatically'. This worried pilots who thought they might be out of a job and so they turned on it: teasing it, calling it names and taking its dinner money. The One Thousand and Eleven was OK: nothing majorly happened – the Japanese prime minister was caught up in a bribery scandal over his country buying the jet, but that was about it. Oh and the one that crashed in Florida produced a ghost that appeared on another flight. But that was about it.

The wide body still exists and has gotten bigger. The first passenger-carrying aircraft could carry one passenger and now the biggest one can carry 853 times this amount (853 passengers). It is the Airbus A380, known as the 'double jumbo' by us as that's what it is: the first double-decker airliner. Sadly, passengers are not welcomed aboard with the theme tune from 1970s children's TV show *Here Come the Double Deckers* as that would definitely get them in an upbeat holiday mood.

BOEING V AIRBUS

There have been many aerial battles but the battle of the airliner giants has all the ingredients of a binge-worthy boxset drama. OK, it might not be that one about the police being in the line of duty or that one about the passengers on the airliner who get lost, but it could still pass an hour or two. The tale was full of intrigue, drama and planes – what else do you need?

The set-up was like this: by the 1970s Boeing was selling tons of airliners: the 727, 737, 747 and, to keep the number sequence going, was bringing in the 757 and 767. They were the kings and queens of the long-haul airliner world with the 747 the top dog.

Over the Atlantic, European manufacturers wanted some of this action and so set up a company called AIRBUS (Airliners In Reality Boeing USA Smelly). It was a joint effort by France, Britain, France, Spain, the Netherlands, West Germany and France.

The first aircraft they made was the Airbus A300. After a few years, Airbuses proved they were not like land buses, because they were not coming in threes; they were coming in their hundreds and eventually thousands. Soon they had a market share of about 50 per cent. The company was making the A310 and then the A320. Would there be an A330? Yes. And a 340 and a 350 and a 380. Wait, no 360 or 370? Seems not. They were skipped but you never know, there's still time.

Boeing weren't overly chuffed with this new kid on the airliner manufacturing block. But they couldn't really do much about it. Some in the company suggested taking those B-52 bombers they had made and bombing those pesky factories in Europe, but this was rejected as the pilots didn't want to jeopardise their chances of getting an airliner pilot job after their air force career. The company continued making planes and continued the number sequence with the 777 and 787. The company ran into a bit of trouble when two of its 737 MAX airplanes crashed. As we know, this is never a good thing for the ole' PR.

The rivalry continues, with both companies vying for sales. Why can't they get on? Maybe they should think about merging, then they'd really dominate the market. Imagine the planes Boebus (or Airing) could make? Something to think about, guys.

TWELVE

THINGS WE ALMOST FORGOT OR DRONING ON AND ON

Aviation is so densely packed with interesting aircraft, events and characters, it's sometimes easy to forget some of them. But not us!

HOVERING ABOUT

Although merely functional and nothing great to look at, helicopters can be really interesting. Really. They are not just there to fly hedge fund managers to and from golf courses, but can provide a way of filming fleeing sports superstars or searching for missing shopping trolleys.

Of course, without helicopters there would be no TV shows like *Airwolf, Blue Thunder, Whirlybirds, Helicopter Heroes, Helicopter Heroes Down Under, Rescue, Sky Cops, Helicopter ER, Emergency Helicopter Medics, Helicopter Pursuits: Caught On Camera, Highland Emergency* and the tense, locational 1980s game show featuring Anneka Rice that was *Treasure Hunt*. Of course, no one who saw Hawaiian private detective show *Magnum PI* could forget seeing that colourfully striped Hughes 500D and Tom Selleck's moustache.

But all this was in the future when back at the start of the twentieth century the first helicoptering devices lifted off. In 1901 the German inventor Hermann 'German' Ganswindt (beard) was reported to have created a machine that lifted two humans off the ground for a short time. There was said to be film footage of the flight but it is no longer able to be located, but they said there was film of the Roswell alien autopsy and we know how *that* turned out.

A few years later another rotary device lifted off. It was designed by two brothers (obviously): Monsieur Bréguet (full, trimmed) and Monsieur Bréguet (also full, trimmed) along with a professor called Charles 'Professor' Richet (the full bushy).

To our eyes it doesn't look like a proper helicopter but we're not sure what it does look like. A massive drying frame? An agricultural threshing machine? An explosion in a blind factory? The Bréguet–Richet Gyroplane Number 1 gained the name 'quad-rotor', not because it was the size of a university quad but because it had four rotors – each over 26ft in diameter. Its thirty-two blades were to give it enough oomph to lift a pilot off the ground and that pilot was a Monsieur Volumard, who drew the short straw and was encouraged into the cockpit on a famous date in 1907. With the engine at full coal, the machine rose majestically to a height reckoned by those observing to be well over 60cm. It was difficult for these observers to gauge the height exactly, as they were hiding behind a sofa throughout, worried about the spinning blades performing an ad hoc headectomy. The Bréguets gave up on vertical-lift machines for a bit to make proper aeroplanes with wings but would return years later. If we remember to include them.

A more successful vertical-taking-off design was by Paul Cornu (full and bushy). His machine would have come in handy if an attacking swarm of giant flies appeared, for it was fitted with six large swatting aerofoils. They worked though and he flew for twenty seconds. OK, it wasn't utterly amazing but it showed the general principles. In many aviation history books, Cornu is regarded as achieving the first controlled flight. Pauly passed on progressing his project when he couldn't get cash but remained an advocate of vertical flight. He was tragically killed by aerial means just before D-Day in June 1944 when his house was hit by Allied bombing. Of all the ways for an aviation pioneer to go, this was unique.

A machine that looks like a helicopter but isn't one is an autogiro. Devised by Spanish person Juan 'de la' Cierva, the 'autogiro' (or 'autogyro' depending on which online dictionary you prefer) operates on the principle of a craft being propelled forward by a normal propeller, so providing a slipstream that turns blades mounted above the fuselage, so providing lift with a rear tail-mounted rudder correcting any torque.* The autogiro/gyro didn't become hugely popular but the insights gained helped the development of the helicopter, so that was good. Sadly, Cierva was unable to be personally involved in this, as he died in a plane crash in 1936. The autogiro/gyro that did become hugely popular was *Little Nelly*, due to its appearance in the James Bond film *You're Only Young Once*. It was fitted with fake rockets and machine guns and was a big favourite at air shows, flown by Ken 'William' Wallis.

The 1930s saw much activity in several countries to build a proper helicopter. Monsieur Bréguet** was back with his Gyroplane Laboratoire (Gyroplane Laboratory), which had two rotors spinning around, and is generally taken by those who know their stuff to be the first practical helicopter design. Showing that lightning can strike twice, the Gyroplane

* **If you think that was hard to read, you should try writing it.**
** **We remembered!**

prototype was smashed to bits by Allied bombers doing their bit to destroy the French aero industry. We do not know if the pilots went on to fly for Boeing ...

Vying for the title of first functioning helicopter was the twin-rotor Focke-Wulf Fw 61. As its name suggests, it was flown in Nazi Germany by Nazi pilot Hanna 'Nazi' Reitsch.* At the Deutschlandhalle (Deutschlandhalle) in 1938, Hanna flew it *inside* the hall. The audience were wowed by this new innovation, although some would have preferred more demonstrations of how to smash up Jewish shopkeepers' windows.

Sikorsky will be a familiar name to those paying attention as it is that of Igor 'Russian Designer' Sikorsky, who had come up with those massive planes years before. His VS-300 helicopter looked like something we would recognise as a helicopter, with big spinny rotors on top and a little spinny rotor at the back. It achieved a notable first as the first helicopter to land on water – not by accident but by design, as the machine had big floats bolted on to the fuselage. Nice one, Igor.

Sikorsky flight tested his own machines wearing his trademark fedora. You'd think the strong winds right above

* **Reitsch test-piloted many Nazi planes, including jets and rocket-powered aircraft. To prove she was very keen on her Führer, she volunteered to lead a crack(pot) team of suicide pilots, which even the diehard friends of Adolf blanched at. In 1945, as the Soviet Red Army and the end of the war approached, Hanna flew into Berlin to see her idol in his bunker. Most who come back from Berlin have a fridge magnet or a tea towel, but Hanna returned with a cyanide capsule. There were rumours she had also come out with Herr Hitler. Did she allow him an escape route to South America where he could lead other ex-pat Nazis to plan their return by doing ad hoc dental procedures? Probably not.**

his noggin would see the hat fly off, so maybe he used superglue. We just don't know. His helicopters were procured by the military, who painted them green and called them R-4s. They were used in the Second Big War, something we didn't know until recently.

KEEP OFF – PRIVATE JETS

Whether it's a hand-carried sedan chair or an internal-combustioned limousine, the wealthy have always had more comfortable modes of travel. With aeroplanes, the rich get to travel up the front, get wider seats, sip chilled champagne and use ermine ear plugs. But in the 1960s the well-heeled had a new mode of getting about when private jets came on sale. Some of these were Learjets, developed by William 'Lear' Jet. They were about the size of a fighter jet but instead of missiles they carried six passengers.

Other private jets were made too, like the Lockheed JetStar, the Hawker-Siddeley HS.125 and the French Falcon, and celebrities queued up to be seen flying in these cool new ultimate status symbols. Frank Sinatra bought a Learjet to fly him and his Flat Pack buddies, Dean 'Aston' Martin and Sammy 'Davis' Jr, around. Elvis acquired a JetStar. Businesspersons wanted in on the act too, and so the 'bizjet' (short for 'bizzjet') was born. This allowed high-ranking chief executive officers to claim it was essential travel rather than slum it with the other proles in business class.

Like Minis, waist sizes and Britain's national debt, private jets have got bigger over the years. A Bombardier Global Express 6500 is just 7ft shorter than the company's CRJ700 airliner, which carries sixty more people. The Gulfstream G700 can fly almost 9,000 miles in one go. That may seem a long way, but its walnut interior and white leather reclining seats will make it feel like a trip to the shops. Shops that sell unicorn pelts. Comfort doesn't come cheap, unless you think $75 million is cheap.

But if this isn't expensive enough, why not look at a converted airliner? Normally packed with hundreds, they can be just for you, because you deserve it. It's now possible for one person to fly about to their heart's content in a:

- Boeing 707 (John 'Dancer' Travolta)
- Boeing 747 (Sultan 'of' Brunei)
- Airbus A380 (Prince 'Al-Waleed bin' Talal).

Sadly, private jets haven't always provided a safe mode of travel. Formula One driver David 'Driver' Coulthard survived a Learjet crash, while golfer Payne Stewart didn't survive a Learjet cabin depressurisation. Flying is risky but if you're going to go, going in comfort is maybe a bit better.

AIR FORCE ONES

With celebs and business leaders swanning about in private aircraft, it was no surprise when politicians wanted to swank about a bit too. Nothing says 'Hello, here I am' than a very big jet with a big flag on the tail. American presidents have two Boeing Air Force Ones, although one was lost when President 'James' Marshall ran into terrorists loyal to General 'Ivan' Radek while returning from Moscow.*

British prime ministers always felt a little bit ... smaller when they had to stand at an airport and watch their American counterparts get out of a big airliner. For years they had to

* **You sure? Seems similar to the 1997 Harrison Ford movie *Air Force One*? – Ed.**

make do with priority boarding on e***Jet. This rankled. They needed to be flying a flag and so in 2020 Britain got its own big jet when a European-built Airbus A330 Voyager tanker/transporter was painted up like Noel Gallagher's guitar. Another, smaller, Airbus was given the same paint job but unfortunately, due to the pandemic at the time, there was nowhere exotic to go, so the prime minister flew to Cornwall.

Going back a bit, one man who had his own private jet on call was Howard 'Enigma' Hughes. In the pantheon of eccentric aviation entrepreneurs who ended up reclusive figures holed up in hotel rooms surrounded by their own urine, Hughes comes top of the list. He made movies, a fortune and aircraft. The most famous was the H-4, given the nickname 'Spruce Goose' as it was an airplane made of wood, not actually spruce but birch. Nothing really rhymes with that.

The big flying boat was the biggest plane of its day – and many afterwards – but only flew once before becoming a visitor attraction. Hughes receded from public life and became very eccentric, shunning publicity and washing. Appropriately for someone who liked airplanes, Hughes died in one (a Learjet, of course) while going to hospital. His life was enough to make your toenails curl.

TEAMING UP

It should be remembered, especially by us, that not all military flying is about dropping missiles or firing bombs, or the other way round if they're doing it right. If budgets are available, aerobatic teams can show taxpayers the skills of the aviators they are paying mightily for. America has the Blue Angels and the Thunderbirds, while Canada has the Snowbirds. The American teams use front-line fighter jets such as F-18s and F-16s, while Canada's use Canadair Tutors. These are quite old, dating back to Roman times.

Not quite as old are the jets used by the RAF's current display team, the Red Arrows. They fly Hawk jets, usually nine of them, unless there's a good film on TV and then an eight-ship will go up. Before the Arrows formed, Britain had teams such as the Pelicans, the Shandies, the Jigsaw Pair, the Yellow Fevers and the Dave Clark Five.

FINDING SPACE

In 1961 an epic flight took place that would echo down the decades and probably the centuries. This flight was different from every one before it, as it wasn't in the air. It was in the space. The Outer Space.

The flight was paid for by the USSR, who had built a space rocket for this very purpose. The first cosmonaut* was called Yuri 'Cosmonaut' Gagarin and with a blinding blaze of flame, but not publicity (the Soviets were not keen on this sort of thing), Gagarin's *Vostok One* lifted off the launch pad in a secret location somewhere in the USSR and didn't blow up, unlike early American rockets. The rocket kept going and going, upwards into the wild blue yonder and then the darker blue yonder and then the blackness of space yonder. Cosmonaut Gagarin was in orbit. He'd done it! His place in the *Vodka Book of Records* was secured.

Gagarin's enthusiasm and joy was clear and he enthralled listeners with his descriptions of the Earth: 'It's very big,' and the spacecraft: 'It's working.' This joy wasn't replicated in the US of A. Those working on the space program** had really,

* **Space term for 'pilot'.**
** **American word for 'programme'.**

really wanted to be first. The American public were aghast and blamed the 'goddamn communists'. They were right. It was indeed the communists who were first. The Soviets had been first in other space ways: the first to put an inanimate object in space (Sputnik the satellite) and an animated object (Laika the dog, who sadly became unanimated).

This was, of course, the Space Race and one big prize lay ahead: The Who Was Going To Be First To Land A Human On The Moon And Return Them Safely To Earth Prize. This would trump all the other firsts, unless it was the Soviets winning it, in which case the Americans would say they weren't that bothered in the first place before stomping off to the basement for a proper sulk. They really wanted to be first and were going to give it their best (Moon) shot.

The might of American industry, government money and ex-Nazi rocket engineers went to work. In 1961 President John 'F' Kennedy stated the country's aims: they would send a person to the Moon and if they behaved, they would be brought back. This was to be done by the end of the 1960s to keep everything nice and neat for historians. And if it happened to be done within JFK's presidency then that would be fine and yankee doodle dandy too. 'What, we've only got two years to do it?' asked Dallas's Mystic Meg.

Without repeating much of what is available in that fine book *Project Apollo: The Moon Odyssey Explained*, published by The History Press and available in all good off- and online booksellers, the Americans sent several rockets towards the Moon. These cramped and smelly tin cans carried men with names like Al, Buzz, Ed and Pete. To relieve the monotony, they joshed, flicked each other with wet towels, put itching powder in their space shorts, told crew members there was an explosion in the oxygen tanks – good old-fashioned bantering stuff like that. The Soviets had tried to build a rocket to take their cosmonauts to the Moon, but they kept blowing up. Each time they watched a US rocket blast off, there was a collective gnashing of teeth.

The gnashing reached a crescendo on a famous date in 1969 as two astronauts* landed on the Moon in their spaceship, *Eagle*. Having come all that way, it would be churlish not to have a look. The first down the spaceship's ladder was a person whose name would be forever etched into the history books, never to be forgotten. ~~Nigel~~ Neil Armstrong reached the bottom of the ladder, then uttered the immortal words, 'I definitely turned the gas off, didn't I?' He had.

Soon, ~~Edward~~ Edwin 'Buzz' Aldrin made his way down the ladder to immortality. A few hours later, after taking photographs and stealing lunar stones (despite the National Parks notice 'Leave only footprints'), they clambered back into *Eagle*, with Buzz pondering this epic moment: when was he going to put the itching powder in Nigel's suit?

The Americans took a break from space flying as they had a war to pay for, but in 1981 they were back with the Space Shuttle, which only suffered two catastrophic incidents leading to the deaths of all on board. Unfortunately for NASA, they were on TV.

MORE EPIC FLIGHTS

Although the glory days of epic flights were mostly in the past, occasionally someone would dream up a chance of glory. One such flight was conjured up by that traditional avenue: brothers – the Rutan brothers, Bert and Ernie** Rutan. Bert designed a one-off aircraft he called *Voyager* and this very thin but elegantly balanced twin-boomed long-winged beauty was to

* **American word for 'cosmonaut'.**

** **Dick. – Ed.**

go on a very epic voyage indeed: a flight around the world. 'So what?' you might say. 'Hadn't that been done loads of times?' And it had. But not in an airplane called *Voyager*. On a famous date in 1986 Dick and Jeana 'Not Chuck' Yeager clambered aboard. They started off down the runway at Edwards Air Force Base and kept on this very long runway for a very long time until eventually lifting off. They wouldn't return for nine days. Nine days that saw the couple fly non-stop without refuelling. They'd done it! 'Do we get a commemorative tea towel?' they asked. 'A student accommodation block named after us?' No, they did not. Those days were gone.

Another not-quite-as-epic-but-worth-mentioning flight had taken place a few years before when Bryan 'Pedaller' Allen became the first person to pedal an aircraft over the English Channel. The *Gossamer Albatross* was so named because of its light weight and propensity for following trawlers looking for fish. The flight took almost three hours. The plane could have gotten over quicker but those mackerel sure are tasty.

WHAT, NO ENGINE?

Aviation started with humans dreaming of being up there with the birds. Nowadays that real proper flying experience can still be had. You can jump off a mountainside wearing a 'wingsuit' designed to make you resemble a flying squirrel (the suits have pockets for your nuts). You can strap yourself into a sleeping bag and take off in a hang-glider, and hopefully return. But the classic is that done by the pioneers: gliding.

Young air cadets get their first taste of ~~fear~~ flying in a glider. The young burgeoning aviator grips their parachute straps so tightly their knuckles become see-through as a mechanical winch hauls them up into the sky at an angle no human should experience unless in a theme park and goaded into it by their bullying mates. Once up, the glider begins its remorseless descent to the hard ground that comes up pretty quickly, especially for those cadets who forgot to close their eyes.*

In the right hands, modern gliders can really soar. Santa 'Klaus' Ohlmann flew more than 1,800 miles over South America in 2003. A good pilot or one scared of landing? Actually, he was a good pilot. He etched his own place in the *Gliding Book of Records* when he was first to fly over Mount Everest. One of the key things about flying a glider is keeping an eye out for a landing site in case you lose all your lift and you have to come down. The Himalayas are not known for their large flat areas of grass but luckily Ohlmann didn't have to convert his glider to a sledge and landed safely. Well done, Klaus!

EVEN MORE EPIC FLIGHTS

To many modern-day aviation enthusiasts, balloons shouldn't be seen outside children's parties, but they have made epic flights decades after making their first ones back in Chapter Two.

One was in 1999 when a balloon with the product-placing name 'Breitling Orbiter 3' lifted off from Switzerland. It contained helium, hot air, Bertrand 'Jean-Luc' Picard, Brian 'Brian' Jones and plenty of air freshener. Their aim was to float around the world. Sounds easy enough when it's put like that, but it's actually quite tricky. B & B hoped to use the 'jet stream', which is a type of non-gastric wind that blows off around the world. If the pair could tap into this wind, they could move right round the globe and become the first in a

* **Like us.**

balloon to do so. Without going into unnecessary ~~research~~ detail, the two bold balloonists landed in the Sahara Desert twenty days later. They'd done it!

The pair had been rivalled in their quest by several other attempts. One was by publicity-shy record label owner Richard 'Humble' Branson. In 1987, with pilot Per 'Diem' Lindstrand, Richy made the first hot-air balloon crossing of the Atlantic. When it looked like they were going to land in Scotland, the pair jumped into the sea. They then flew over the Pacific a few years later, reaching speeds of 245mph. That's the sort of speed that would make your beard flap up over your face.

When Rich turned his mind to circumnavigating* the globe, it wasn't plain sailing. He had three goes:

1. In 1997 his balloon lifted off from Morocco. It lost helium and height and put Branson in a pickle. He decided that there was no point continuing, so landed in Algeria.

2. The same year another attempt was mounted. Before he could get in it, Branson's balloon broke free and floated up to 60,000ft and headed off east. Bye, balloon, don't forget to write.

3. A year later his balloon got away okay from Morocco with the dashing aeronaut inside this time. It managed a good bit before ditching in the Pacific near Hawaii. The brave Sir Richard consoled himself with having lots of money.

* **Navigational term for 'going around'.**

BAD THINGS

Aviation, as we've probably mentioned too many times,* carries inherent risks. Safety has improved massively as aircraft developed – at least for airline travel rather than for hang-gliders.** You could fly all your life without encountering any Bad Things, save for injuries sustained in the Battle of the Overhead Lockers.

But Bad Things do happen. Major events that made it onto the TV news included Tenerife in 1977, when two Boeing 747s collided; Sioux Falls, where a DC-10 crew flew without any operating control surfaces; and New York, when Captain 'Sully' Sullenberger tested the flotation capabilities of an Airbus A320. The risks increased greatly when Bad People got involved. Terrorists saw aeroplanes as suitable vehicles for carrying their causes, and from the 1960s onwards it was difficult to board a flight without having to queue up behind Red Meinhoff Liberation Faction Front members. They were keen to hijack airliners and get them to fly where they wanted rather than what was printed on the tickets. What was so great about Cuba?

In one famous incident a trio of airliners was blown up by one of these 1970s gangs. Quite what they were doing it for is lost but it must have been very important. The sight of a VC10, DC-8 and Boeing 707 being turned to scrap metal in the desert was a shocking sight, especially for those whose duty free was still on board.

Then at the start of the twenty-first century, terrorists used aircraft in a way that isn't great material for comedy. American airliners were used to hit large buildings in

* **You think? – Ed.**

** **The quickest way to get £250 from Harry Hill. Or a payout from your life insurance provider.**

America, causing the deaths of thousands of people who were at their work. Many thousands of words in reports, articles and books have been written about this event but after twenty years or so they are still shocking and bewildering. A sneakier way was to plant explosives in a plane filled with civilians with little gripe against anyone and then blow them up. It happened enough times to make some doubt the words 'civilisation' and 'humanity'. Said it wasn't great for comedy.

UNPERSONNED AERIAL VEHICLES

You're on a quiet country walk or maybe strolling across a beach at sunset. You hear a buzzing noise. Is it a swarm of angry wasps after your jam sandwich? Well, it might be, in which case you better start scoffing that butty pretty quickly. But more likely it is the curse of the reflective moment: the Bloody Drone. Drones are small flying machines that use tiny propellers to annoy everyone within hearing distance – apart from the proud operators, chuffed to have kept their new toy within sight and sound. Drones closed Gatwick airport in December 2018 as they were spotted flying into the airport's airspace. Security staff carried out their standard countermeasures: swearing while waving their fists, but to no avail. In order to avoid drone-to-airliner ~~collisions~~ incidents, the airport was closed. The drone operators were never found and it's maybe just as well. If they fell into the hands of those families being denied their Christmas break, they might have had difficulty retrieving their control yokes.

There is another type of drone, although you are unlikely to see one unless you are planning an IED attack in a hot and dusty country. Military drones do a lot more than annoy people, being able to drop missiles or bombs. They have several advantages over piloted aircraft in that, if shot down, there is no collateral* to be captured. They can also stay in the air for much longer, not needing to breathe oxygen or go to the toilet. Yes, there is a risk they can drop their weapons on the wrong people at a wedding or birthday party and not give a monkeys but what is life without risk? We've all got to go sometime, yeah?**

BRITISH MADE

Ha! You thought we'd forgotten about this.

The merging of British aircraft companies had continued until – to quote comical fantasy movie *Highlander* – there could be only one big one. That one is BAE Systems. It no longer makes civil aeroplanes or bits of them. That ended when it sold off its share of the Airbus money tree. When it came to making airliners, Britain had decided to go out on a high. The last-ever British airliner, the Hawker Siddeley-British Aerospace-BAE Systems-Avro RJ 146X, was one of the most successful the country had built, selling hundreds of them. The company philosophy was, why should they continue making planes, when they could stop and then not make any money? Yes, that's what we thought.

BAE still made military planes – that was a money tree no self-respecting accountant could turn their back on. Do you know how much you can charge for a camouflaged spanner? Millions! A lot of the planes were jointly built by foreign countries. The Jaguar (UK and France), Tornado (UK, Germany and Italy), Typhoon (UK, Germany, Italy and Spain) and Hawk (Scotland, Wales, Northern Ireland and England). These planes weren't cheap but much of

* **Military term for 'human'.**

** **No. – Ed.**

the costs came from translation services. What is Italian for 'missile'?

AIRSHIPS AGAIN

In the twenty-first century, planespotters and undertakers were agog, awed and amazed to see an unfamiliar shape return to the skies. Yes, it was an airship! The Airlander 10, in fact. A team of designers, engineers, HR practitioners, receptionists, administrators, IT technicians and project management staff, among others, had worked for several years and now their project was in the air. They then saw it immediately nicknamed 'The Flying Bum', due to its twin-cheek design. That's harsh.

Measuring 300ft long, it had been intended for the US Army, but they didn't want it. Instead, it was developed as a civilian craft able to do various things. Unfortunately one of these was crash. As if in a children's book called *The Flying Bum Who Wanted to Fly*, one day it broke free of its moorings. Desperate to float with the clouds, the Balloony Backside made for the dreamy blue skies. Its flight was short-lived as the automatic deflation system operated, producing a farty noise heard more than 30 miles away. The Airlander was written off, but it'll be back, in the form of Airlander 50. What is it with number sequences? Was 11 no use?

THE FUTURE

In the history of aviation there have been many turbulent bumps along the way. War, pestilence, famine and French air traffic controllers are just some of them. With concerns about aircraft and their emissions' effects on planet Earth, and all of the carbon-based life that lives on it, it's hard to predict the future for powered flight. Electric planes? Gas-powered planes? Or just sit on the ground and pick the belly button fluff out while watching videos of planes from the past?

No. The human desire to fly will always remain strong. The draw of being able to rise up and look down on the ground that we were on only a few minutes ago will be one that continues to enthral and pull us up out of our ... or we could just get a drone with a camera and watch that.

ACKNOWLEDGEMENTS AND SOURCES

Many thanks go to Amy Rigg, Jezz Palmer and the staff at The History Press for the opportunity and for their unfailing professionalism in the face of such bad jokes and puns.

SOURCES

Although it might not look it, some research was done to check what really happened through the decades when epic aviation history was being etched.

These were consulted but should not be held responsible in any way:

Airshipwreck by Len Deighton and Arnold Schwartzman (Book Club Associates, 1978).
Aviation: The Pioneer Years researched and edited by Ben Mackworth-Praed (Studio Editions, 1990).
Flight: 100 Years of Aviation by RG Grant (DK, 2004).
Reaching for the Skies: The Adventure of Flight by Ivan Rendall (BBC, 1988).
The Great Atlantic Air Race by Gavin Will (The O'Brien Press, 2009).
The Story of Flight (Ladybird, 1960).
The Spectacle of Flight: Aviation and the Western Imagination, 1920-1950 by Robert Wohl (Yale University Press, 2005).
Wings and Space by John Chaplin (Ian Allan, 1970).
Wikipedia.
My dad.

BY THE SAME AUTHOR

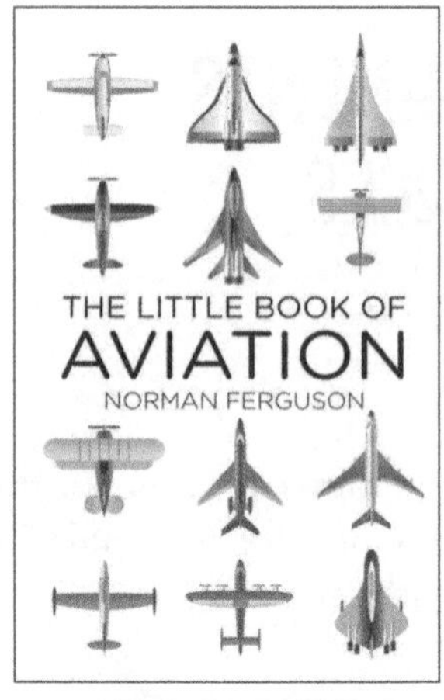

978-0-7524-8837-0

978-0-7509-6838-6

978-0-7509-8978-7

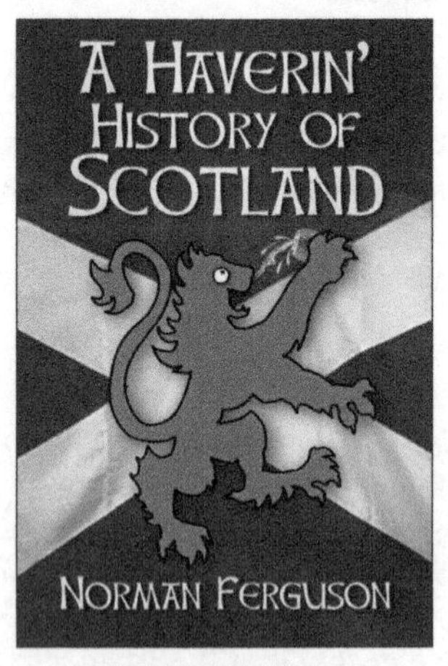

978-0-7509-8693-9